D1104364

Climate and the Affairs of Men

Harper's
MAGAZINE PRESS

Climate
and the
Affairs of Men

Nels Winkless III
and Iben Browning

HARPER'S MAGAZINE PRESS
Published in Association with Harper & Row
New York

CLIMATE AND THE AFFAIRS OF MEN. Copyright © 1975 by Nels Winkless III and Iben Browning. All rights reserved. Printed in the United States of America. No part of this book may be used or reproduced in any manner whatsoever without written permission except in the case of brief quotations embodied in critical articles and reviews. For information address Harper & Row, Publishers, Inc., 10 East 53rd Street, New York, N.Y. 10022. Published simultaneously in Canada by Fitzhenry & Whiteside Limited, Toronto.

FIRST EDITION

Library of Congress Cataloging in Publication Data

Winkless, Nels.
  Climate and the affairs of men.
  Includes bibliographical references.
  1. Climatology. 2. Paleoclimatology. 3. Volcanoes.
4. Man—Influence of climate. I. Browning, Iben, joint
author. II. Title.
QC815.2.W55   551.6   74-27305

ISBN 0-06-129550-7

75 76 77 78 79 10 9 8 7 6 5 4 3 2 1

To Ed Dougherty,
to Peter Johnson,
and to Francis Mair,
for different reasons

There is a tide in the affairs of men . . .

—Shakespeare

A Prologue

We speak of remote places and ancient events in this book, things that in ordinary context seem to have no practical significance for our daily lives. To the regret of the authors, businessmen and government officials have found it difficult to see any short-term value in this work—they've all been too busy putting out fires in their wastebaskets to notice that the building is burning down around them.

In fact, evidence streams past us constantly, indicating the immediate importance of changing climate in the affairs of men. It takes a practiced eye to select appropriate items from the newspapers and magazines, but relevant, currently significant materials appear in remarkable profusion. Every year, for example, China brags up its wheat crop in an effort to depress the market price so they can buy cheaply from Canada and Australia. Congress is scolded for failing to control the level of the Great Salt Lake. The Third World attacks America for promoting population control measures and then demands free food to support its teeming population. Russia tends to fire ministers of agriculture, a great advance over the good old days. Recall that during the bad food-growing years of 1936–38 Russia executed several million Kulak farmers for "underproduction."

Examine this miscellany of news items. They'll put the remote places and ancient events in perspective before you plunge into the text.

HAWAIIAN QUAKE MYSTERY—IONOSPHERE WENT SILENT

. . . the ionospheric blanket—at least above the Oahu stations—
had disappeared. *San Jose Mercury-News,*
September 9, 1973. Byline
Frank Carey, AP.

SKY GLOW "PREDICTS" QUAKES

. . . Valentin Ulomov, deputy director of the Seismology Institute
of Uzbekistan, told Tass the electrons coming from the earth's
interior change the distribution of electric charges in the atmosphere
and cause the glow. . . . Associated Press,
November 8, 1973

VICTIMS OF QUAKE BURIED IN MEXICO

. . . The Mexican quake, worst ever in that country's history . . .
. . . Other quakes struck Thursday in Andes mountains along the
Colombian-Venezuelan border, California's Imperial Valley, and
Alaska.
. . . Seismologists said there were no connections between any of
the week's quakes. . . . *Albuquerque Tribune,*
September 1, 1973 (UPI).

QUAKE SHAKES EAST COAST

A rare earthquake . . . surprised and frightened sleepy residents
early Wednesday . . . Trenton, New Jersey, to Baltimore. . . .
. . . said there was "no way the quake could be related" to a more
intense earthquake that struck the sparsely inhabited group of Soviet
Union Kurile Islands in the Northern Pacific earlier Wednesday. . . .
An "earthquake swarm" gently shaking northern Nevada . . .
reported Wednesday . . . *Dallas Morning News,*
March 1, 1973 (UPI).

QUAKES IN 1972 KILLED 10,000

Denver Post,
February 25, 1973.

AFRICA DUST BLURS SKY OF CARIBBEAN

Dust from a severe and prolonged drought in Africa has been
blowing across the Atlantic for the past five years and is polluting
the air over the Caribbean. . . . *Denver Post,*
February 10, 1974 (AP).

EVACUATE IN PATH OF GLACIER

Soviet authorities have begun evacuating inhabitants of a remote central Asian valley threatened by floods caused by a rapidly advancing mountain glacier.

Albuquerque Tribune,
May 31, 1973 (*New York Times* News Service).

FAST MOVING GLACIER PROBLEM

What is perhaps the largest glacier in mainland Canada has begun advancing rapidly . . . serious threat of floods downstream. . . .

Albuquerque Tribune,
December 6, 1973
(Ottawa Enterprise Science News).

GROWING ICE CAP MAY BE TREND

Weather satellites sweeping across the Northern Hemisphere have come up with a surprise: The permanent snow and ice cap has increased sharply.

. . . increased by 12 per cent in the Northern Hemisphere in 1971 and has remained at the new level.

Albuquerque Journal,
May 15, 1974 (AP).

GREAT SALT LAKE AT HIGHEST LEVEL

Great Salt Lake has reached its highest level since 1929. . . .

Albuquerque Tribune,
May 20, 1974 (AP).

LILACS ARE CONFUSED

A mild autumn apparently confused some growing things on the island of Corsica. Lilacs, cherry trees and apple trees, which normally bud in May, blossomed in late October and early November.

Albuquerque Journal,
November 7, 1973 (AP).

THE GROWING THREAT OF WORLD FAMINE

. . . the Food for Peace program—instead of being *increased,* which is the clear need—may actually be *decreased* 40% from last year's rice quantities and 66% from last year's wheat quantities.

Wall Street Journal,
September 14, 1973.
Roy L. Prosterman.

3

RUNNING OUT OF FOOD

The U.S. now views agricultural products not as a giveaway item, but as a way of earning the foreign exchange needed to pay for imports, including high-priced crude oil. "Food for Crude" is the shorthand for the current policy at the Department of Agriculture.

Newsweek, April 1, 1974.

BUTZ SAYS U.S. CAN'T, SHOULDN'T FEED WORLD

. . . the massive American surpluses are over. . . . In the beginning of the American food aid program, they explained, the food giveaway was simply a device to dispose of the produce. In other words, it was cheaper to give it away than to store it.

Albuquerque Journal,
September 5, 1974 (AP).

DROUGHT THREATENS CHINA'S FOOD

This is the third dry season in succession on the North China plain.

Albuquerque Journal,
June 29, 1973.

AUSTRALIA SELLS WHEAT TO CHINA

. . . more than $900 million at current international prices . . .
. . . the first time that Australia has negotiated a long-term wheat agreement with China.

Albuquerque Journal,
October 12, 1973 (UPI).

WHY THE RUSSIANS DO WHAT THEY DO

Russia has an even more basic reason for turning westward: food. Because of frost damage in the Ukraine and other areas, the USSR expects an exceptionally poor harvest of winter wheat this year. It needs the pending wheat sales from the U.S., the largest since the cold war began, to help feed its people during the next year.

Time, May 22, 1972.

RUSSIANS FIRE MINISTER IN AGRICULTURE CRISIS

. . . responsible for the worst farm crisis of the century . . .

Denver Post, February 4, (UPI).

DEATH TOLL CLIMBING IN ETHIOPIA

Between 50,000 and 100,000 persons have died since April in a drought . . .

Albuquerque Journal,
October 31, 1973 (UPI).

ETHIOPIA: THE LION AT BAY

After 43 years of one-man rule, the 81-year-old Selassie—the Conquering Lion of the Tribe of Judah, the King of Kings and the Elect of God—was no longer in control . . .

. . . ineptitude in dealing with the drought . . .

Newsweek, March 11, 1974.

KING FAMINE

. . . The drought has caused even greater disruption in Upper Volta, where a southward migration of more than a million people is under way. Nomads are pouring into the Ivory Coast and Ghana in a search for grazing lands. Their starving animals are poaching on cropland tended by subsistence farmers. The result has been a number of pitched battles similar to those between cattlemen and sodbusters in America's Old West. *Time,* April 30, 1973.

STUDENTS CELEBRATE COUP IN NIGER

. . . charged over Radio Niamey that Diori has mishandled the "disastrous situation" resulting from the six-year-old West African drought . . . *Albuquerque Journal,*
April 17, 1974 (AP).

CITY OF ETERNAL SPRING SHIVERING AFTER STORM

MEXICO CITY (AP)—This city of so-called "eternal spring" woke shivering Saturday to one of its most drastic temperature changes in 20 years.

. . . caused the temperature to drop from 68 to 38 degrees in twelve hours. . . . *Albuquerque Journal,*
December 17, 1972.

10-MONTH DROUGHT HAS INDIANS TALKING OF FAMINE

. . . Mrs. Gandhi told hungry peasants 10 days ago that the government's relief-work program is the largest ever undertaken to combat famine. *Albuquerque Journal,*
January 11, 1973 (AP).

DROUGHT BRINGING RUIN TO S. AFRICAN FARMERS
Johannesburg, January 5, 1973.

DRIEST SINCE 1749
DROUGHT THREATENING BRITAIN
Denver Post, May 6, 1973.

THE YEAR OF THE FAMINE

. . . In the Philippines a 740,000-ton shortfall forced the government last week to launch an emergency rice production drive . . .

Newsweek, June 4, 1973.

ELEPHANTS HUNGRY

A herd of more than 400 elephants in search of food and water has devastated thousands of acres of cropland this year in southern Chad, according to reports reaching Fort Lamy.

Albuquerque Journal,
June 8, 1973 (AP).

CYPRUS IN GRIP OF WORST DROUGHT SINCE 1881

Denver Post, August 19, 1973.

SOVIET SPRING SOWING DELAYED BY WEATHER

. . . The 1972 crop was described by Soviet officials as a "once in a century" failure. . . .

Albuquerque Journal,
May 10, 1974 (AP).

THE POPULATION BOMB TICKS ON

[Reporting on the U.N. Conference of World Population]

As it turned out, the most basic split was between nations genuinely alarmed by soaring population—and determined to limit it—and those that felt that people are a useful resource.

. . . a roaring baby boom is no obstacle to economic growth . . . blaming poverty on the U.S. and Russia . . .

Newsweek, September 2, 1974.
Byline Anthony Collings.

1

Let us be clear from the outset: the ideas we'll be discussing in this book are not the products of a massive government study in which armies of gowned scholars and exotically equipped scientists were turned loose with unlimited money to unlock the secrets of nature. Instead, this material is the product of a small number of underfinanced people who have been plugging away at a low but steady level for a long time. The big study remains to be done.

We offer here the basis for that study, the observations and conjectures that seem to give us a real grip on man's affairs, accounting for things that *have been* and letting us predict what *will be* with increased accuracy.

In 1963 we were involved in a project that attempted to use active volcanoes as infrared beacons by which to navigate orbiting space vehicles. Up to that time, satellites were all navigated by fixes on stars. Stellar navigation worked just fine. One could tell with great precision where he was with respect to the star Canopis. Unfortunately, he had more difficulty figuring out whether he was flying over Perth, Australia, or Perth Amboy, New Jersey.

The Browning proposal to fix on volcanoes, hot springs, isolated islands or lakes, or even steel mills was greeted with considerable interest and we set out to map a useful grid of such things. Though our client didn't win the contract for the navigation system that eventually went into Skylab and the proposal was transmogrified over the years into something we can't recognize, we did get a chance to do some fascinating work.

A question of significance at the time was "What does *active* mean?" When is a volcano no longer a volcano? Most people are satisfied that a volcano doesn't count any more when it quits throwing hot lava and pumice all over the place. Yet volcanoes do start and stop with some lengthy intervals between eruptions. Indeed, hundreds of years may pass between eruptions of a particular volcano, during which time forests and towns may grow luxuriantly on the mountain slopes. It is often a great surprise and annoyance to people who live on quiescent volanoes to discover that their real estate is exploding beneath them.

We were looking for distinctions somewhat finer than the presence or absence of flying melted rock, and the task was more difficult. For example, is Mount Shasta in northern California a dead volcano? It has been lying there quietly as long as anyone can remember, but John Muir noted in a diary that he had been trapped in a blizzard on Shasta's slopes and had saved his life by curling up next to a nice hot fumarole on the mountain. Similarly, Mount Rainer has suspicious hot stuff at its top and so do a lot of other nice, quiet, respectable mountains.

In the process of tracking down and examining the histories of remote and interesting volcanoes, we began to see some patterns in their performances. More, we began to see relationships between the volcanic activity patterns and other historical events like famines.

We were treated in 1963 to a couple of spectacular volcanic events that stirred our interest further and gave us a chance to make a prediction that could be tested against real future events.

The first event was the sudden appearance off Iceland of a new volcanic island, which established itself so firmly that it is with us still today, and has a name of its own, Surtsey. Surtsey's birth gave us an immediate look at the direct environmental influence of a brand-new volcano.

The second event was the eruption of Mount Agung in Bali. Agung is very close to the equator, not far from Krakatoa. Like Krakatoa, Agung fired a very large amount of dust into the upper atmosphere for distribution around the world by high-altitude winds. We guessed that the effects produced by the dust would reach the latitude of the United States about five years later, seeding increased rainfall for a while and then leaving a drought where the air had been wrung dry of its moisture.

Sure enough, by 1968 we had all-time record snowfall in the Sierras and by 1971 we had severe drought in the Southwest and in Florida. One nice thing about experiments in this field is that you don't have to spend any money setting them up. The volcano goes when it feels like it and you simply collect the results. Patience helps, of course, because the results aren't all in for many years.

Since we couldn't do anything directly to speed the answers to our projections about rainfall, we had plenty of time to look at the histories of similar events in the past.

Others have speculated on the relationship between volcanic activity and good weather for growing crops. Benjamin Franklin wondered in the 1780s if there were any connection between the eruption of Mount Hecla in Iceland and the prevalence of widespread "dry fogs" which were distressing people in Europe. Franklin apparently didn't pursue the matter in detail or he might have discovered that there were simultaneous eruptions on a large scale at various places around the earth, and he would have tied all this together for us almost two hundred years ago.

We turned up a number of periods of such correspondence between volcanic activity and poor food-growing conditions. Famines, of course, correlate with poor food-growing conditions—and wars, bad humor, and upset in the government correlate with famines.

In the course of such study, one is inevitably drawn back to more and more basic factors—the causes behind the causes.

It wasn't long before we had left the concerns of daily weather for the more general subject of *climate*. It wasn't long before we left the examination of individual volcanoes for the study of the periodic rise and fall of volcanic activity worldwide. This led to the study of tidal forces, controlled by astronomical movements, which appear to influence volcanic activity. Astronomical concerns come up again when we deal with sunspots and the well-known proposition that worldwide precipitation correlates with the double sunspot cycle.

We could not avoid history, either: the reports that indicate man's response to changing environment. Over the years we have dealt as much with history as with technology, an odd situation since we are technologists by trade. To some of us, technology has no intrinsic interest but is fascinating only to the degree that people interact with it. A stone ax, for example, is a fairly drab, fragile object that does not hold a practical man's attention for long, unless he is of an inquiring mind and tries to make a sharp stone ax for himself.

A stone ax becomes a lot more interesting after you have spent a few hours finding an appropriate chunk of rock and have then mashed and cut your fingers badly in the effort to shape the stone into a useful object with a sharp edge. Imagining that your life depends on successful manufacture of a good cutting tool, you develop a lot of sympathy and an enormous respect for your ever-so-ancient ancestors who learned to make graceful and effective stone tools in quantity. Those people were interesting.

It turns out that meteorology, climate studies, vulcanology, astronomy, and dendrochronology—disciplines with which we have been dealing to develop our understanding of human affairs—are about as interesting to the average passer-by as a stone ax. People who feel an urge to dig into these matters are tolerated as harmless cranks, but if a television program on climate variability were to appear in place of a scheduled major sporting event, the climatologist in charge would be lucky to escape the enraged viewers with his life.

For the first several years in which we were pressing this work, we were tolerated as harmless cranks, and luckily we had other means of making a living. Indeed, these years were dominated by the folk who talked chiefly about the degree to which man affects his environment. The ecology craze has been in full flower, centering on that rascal man and his wicked abuse of his powers. It was very difficult to distract anybody's attention from man as tyrant over ecology rather than its serf. Far from governing the forces that shape the Earth, man is a virtually powerless creature governed *by* those forces.

Times do change. The decade of the seventies has made the things with which technologists deal more and more interesting as their relationship to people becomes more apparent. Actually, 1973 was the year in which a big change in attitude occurred.

1973 was the year that everybody on Earth really noticed.

1973 was rich in startling experiences.

Europe and Japan discovered that they were oil slaves to the Arabs, having no other significant sources of energy available elsewhere. Even the United States found that it was short of petroleum and would be in an energy bind for perhaps a decade.

In Africa, the monsoons failed for a sixth year and mass starvation overtook a broad band of sub-Sahara countries from Mauritania to Ethiopia. Millions of people were dying. Millions more moved into the marginal lands of their neighbors to compete with them for food.

When a full eclipse of the Sun occurred in the middle of this, one tribe in Africa entered into a debate over what action might be taken to prevent such occurrences in the future. One school favored punishing the Europeans, who had obviously caused all the trouble and capped it with the darkness at noon. A second school favored punishing the king of the tribe, who was, by definition, responsible for everything. The debate was resolved by the general realization that they couldn't get at a significant number of Europeans, while the king was handy. They deposed the king. They haven't had another full eclipse of the Sun since.

In India, the failure of the monsoon and consequent drought assured a massive famine while great floods in Pakistan killed thousands of people. One area in northern India had an unprecedented cold snap. The temperature dropped fifty degrees in what was normally a steamy rain forest. People froze. Fuel supplies were exhausted.

The anchovy fisheries off Peru and Chile failed, apparently because the Humboldt Current was not flowing in the fashion that people had assumed was normal. Since the fish meal was a major Chilean export, this put additional strain on the Chilean economy, which was already involved with expensive major food importation, and a debate broke out over what actions might be taken to prevent the occurrence of hard times in the future. A revolution occurred whose purpose was to depose the country's president, who is, in everybody's final analysis, personally responsible for everything. President Allende hastened the proceedings by dying abruptly of gunshot wounds. The effectiveness of this action has not yet been established.

In the Philippines, floods and generally bad weather put enormous pressure on the economy. By the end of the year a Moslem revolt against the central government was raging, causing a very large number of deaths.

The U.S. dollar took a beating in the money markets of the world and Americans had a hard time figuring out why nobody liked their money any more. Embarrassed, we devalued our currency sharply.

We had other things to think about, too. The Mississippi Valley was flooded in spectacular style and the cotton crop was very badly affected. Farmers planted a late soybean crop when the waters receded because the crop failures in other parts of the world caused a great demand for soybeans and drove the prices up to about four times as much as expected when the year began. We sold so much food overseas that our balance of trade was hugely favorable for the first time in years. Food poured out and money poured in. People here liked the income, but complained loudly about the sharp rise in domestic food costs and the cost of oil-based products. (It costs calories of fuel energy—usually energy derived from petroleum—to produce calories in the form of food.)

Americans were shocked to discover that our food reserves were gone, after two decades during which we had so much that we couldn't get rid of it. At one point the government embargoed export of soybeans and considered an embargo on corn. We heard from Japan instantly on this subject and most of us learned for the first time that a U.S. embargo on corn and soybeans would cut Japan's total food supply by 25 percent. Luckily the soybean crop was good and corn better than expected.

Americans were also annoyed by a realization that they had been slickered in their export dealings with the wily Russians. Russia had its troubles, too. The 1972 grain crop had been a disaster and the Soviet empire was looking forward to hungry times. Russian traders quietly bought grain in the United States, sensibly making many small deals with individual suppliers all over the country. The purchases were huge when added up, and the prices would have shot sky-high if the Russians hadn't kept the matter quiet until after all the deals were made. Well done by them, but it was small satisfaction, compared with the embarrassing spectacle of mighty Russia dependent on its enemies for food. They salved their pride by firing their agricultural leaders and declaring that the 1973 grain crop was a record success. Soreheaded Westerners voiced their suspicions that the claimed 1973 grain tonnages included very large amounts of substandard material that could not be eaten.

China bought massive quantities of grain.

With China and Russia strongly in the market, the underdeveloped countries virtually bankrupted themselves to pay extraordinary high prices for food at the expense of everything else in their economies. Many of their citizens protested this by dying.

In the United States tempers were frayed by a record number of tornadoes. Not only were the twisters very numerous during the summertime, but instead of tapering off gracefully the way they are supposed to by October, they shamelessly kept appearing right on through winter. We'd have been startled to have a thousand tornadoes in a year. We were astonished and indignant to have more than eleven hundred.

Irritated beyond endurance, some Americans began strong efforts to depose the President, certain that in the final analysis, he is responsible for everything.

Everybody hated 1973 and longed to get back to normal. The year was so upsetting that people even turned to studies of meteorology, climatology, vulcanology, dendrochronology, and astronomy, to see if they could figure out what was going on. For the first time in a long time, these matters seemed to have something to do with *people*— and people are interesting, especially when they are close relatives and are starving to death.

Our approach to understanding human behavior involves physical factors that reach far out through space and far back in time. Virtually every proposal we'll make along the way, from basic physical causes to social effects, is controversial, the subject of loud and lengthy fights among experts. Even so, we believe that there is very strong evidence to support the proposals, and the larger part of the disagreement that breaks out is over semantics and detail. When the arguments are substantive, well, that's the way it goes. Here we stand, on the basis of what we have been able to learn and calculate up to now.

Sensible people are often aghast at the tolerances that technical people are willing to accept in their calculations. If you ask a contractor for a rough guess of what it will cost to remodel a room in your home, you will probably be satisfied if he guesses high or low of the actual cost by 50 percent. When you come right down to getting a bid from him, an estimate with which you expect him to live, you'll be upset if the work costs 10 percent more than he estimated and you'll worry about his sanity if it comes to 10 percent less.

In comparison, scientists often operate in an incomprehensible and seemingly irresponsible world of wild guesses. At the very tight end of the scale, technical people in some situations aren't happy unless they have calculated the wavelength of laser light to within an Angstrom, one 254-millionth of an inch, or have figured out whether they have ten or only eight molecules of carbon monoxide among a million molecules of air. At the other end of the scale, though, they may not care whether the universe is five billion years old or twenty billion, as long as it's more than two.

Rough estimates are often used to determine whether you're just in the ballpark or not. You can calculate the age of the Earth from the salinity of the oceans, knowing how long it takes, on the average, for salts to leach out of the land and into the sea. If you then compare that figure with an estimate based on the average rate of erosion of the land combined with the known ancient altitude of the land, you'll be very pleased if the numbers are within a couple of hundred percent of each other. The difference *could* be very much greater, considering all the variables that you cannot calculate and all of the missing information.

It's rather like guessing how many beans are in a glass container in a store window when you can see only the front face of the container. You don't really know how big the container is, or that the beans are all the same size. You do have some clues. The contest operator is not likely to double-cross you by putting a rock in the beans. Containers for beans tend to be symmetrical. Beans from a given batch tend to be very much of a size, and so on. All in all, with no more information to go on, you'll feel satisfied if you estimate it within 40 percent of the actual number. Assuming that you have not bet any money on the matter, you take some comfort in the fact that your estimate was pretty good, that your approach is reasonable, that your answer is at least in the ballpark.

We have that same sort of confidence about our calculations here. Granted that we are short of information and that nature doesn't try to be fair, we get answers that are *more like each other than they are like anything else* when we calculate the same things from different data. It all seems to hang together sensibly.

As we begin the discussion, let us here provide an outline of the overall scheme, a sort of road map through the forest of concepts which will be useful along the way.

To start back at the root of things:

1. Much of what happens in the world is driven by energy that is stored up in the Earth, as contrasted with radiation falling on the surface of the Earth from the Sun, for example. We argue that the

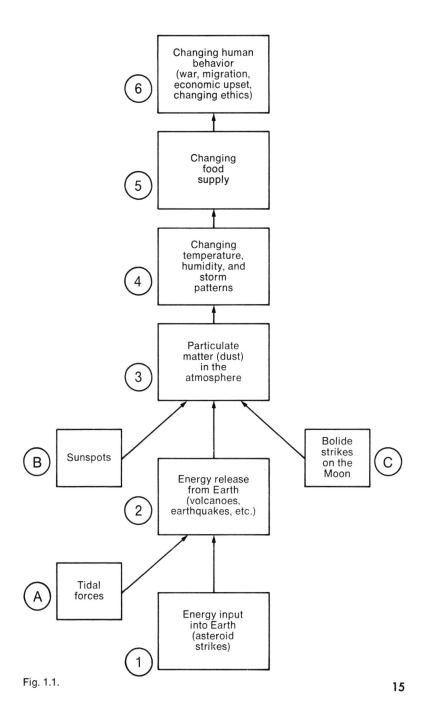

Fig. 1.1.

energy in the Earth is not all just left over from whatever happened at the formation of the planet, but that the Earth gets occasional huge inputs of energy when it is struck heavily by chunks of debris in space, asteroids that become meteorites.

2. The stored energy is released not only as outflowing heat and other radiation, but in earthquakes and volcanoes. We believe that volcanic activity is especially significant in the affairs of men and we'll discuss the question in some depth. Notably:

A. The release of energy by earthquakes and volcanoes is not random, but is very strongly influenced by tidal forces that create stresses in the Earth's crust sufficient to trigger (not to cause, but to *trigger*) quakes and volcanic eruptions. We can very accurately predict the times of high tidal stress, not only to let us know when the momentary peaks occur, but to determine frequency of tidal peaks. These are periods when the Sun, the Earth, and the Moon move close together and line up well to produce frequent high tidal stresses, maintaining a comparatively high level of volcanic activity. At other times, the stresses are less frequent and less severe. We suggest that these periods correlate with human activity and quiescence.

3. Volcanoes fire huge amounts of dust into the upper atmosphere. The dust lingers and is distributed all around the globe. The higher the tidal force and the more frequently tidal peaks occur, the greater is the number of volcanic eruptions and the greater is the amount of particulate material in the upper atmosphere. Note also:

A. Sunspots, which occur in regular cycles, fling huge numbers of charged particles into space and a significant number of these are trapped in the Earth's magnetic field. As they stream from pole to pole they lose energy and become at last "particulate" matter in the Earth's upper atmosphere. And:

B. An appreciable percentage of matter blasted from the surface of the Moon by asteroid and meteoroid strikes falls into the Earth's gravitational field. The Earth is thus liable to have large amounts of dust contributed to its upper atmosphere by more or less random strikes on the moon. This dust contributes to excitement on the Earth as effectively as do volcanic dust and particles from the Sun.

4. The particles in the atmosphere govern climate on the surface of the Earth. They act as a screen to reflect warming sunlight back into space, so that the average temperature of the Earth's surface is lowered. Further, they drift gradually to the surface, seeding rainfall

in the process. Sunspots appear to have a rapid effect on rainfall and one may expect rainstorms in the United States within a week after a major solar eruption (a very controversial notion). The cooling of the surface changes the pattern of weather sharply, shifting the monsoons south so they do not water the south Sahara and India, for example, as they do in warmer times. The effect is worldwide, modified by the terrain and the configuration of the land masses.

5. Changing climate sharply alters the amount and nature of food supply. In a cooling situation, the marginal croplands of the world become submarginal and food production even in comparatively rich agricultural areas is disrupted while farmers adapt their techniques and their choice of crops to altered conditions. Worse, the cooling situation brings with it extreme variation in temperature and humidity which only hardy plants can tolerate. In general, highly productive food plant strains cannot tolerate large variation in environment. Food supplies fall short.

6. People respond to hard times by disposing of their priests, their political leaders, and their excess baggage. War, migration, economic upset, and changing ethics mark hard times. Trouble rolls from the marginal regions along with people as food supplies fail and the folks there move to where food may be found.

The affairs of man appear to be governed, then, by the release of stored energy from the Earth, triggered by extraterrestrial forces. The rate and form of this release are predictable to a useful degree and we suspect that man's activities are thus predictable to a useful degree. The behavior of human societies is surely the most complex matter in this sequence, but it appears to be an effect produced by physical causes.

In the following chapters, we'll work up through this list from the more basic causes to the more general effects, providing documentation of the case at each step.

By the way, our wish to get back to normal is apparently being fulfilled, to the regret of most people in the world. The climate in the middle of this century, which we have taken to be normal, was actually the warmest, most kindly climatic era in many hundreds of years. It was an anomaly, an oddity. The "normal" conditions for life on this planet, the average, are much harder than those under which any of us have lived who now inhabit the earth.

2

In discussing changes in the earth's climate, it is worthwhile to have a good picture of the structure and status of the Earth itself.

Earth is the third planet from the Sun, more massive than either of our near planetary neighbors, Venus and Mars. We are subject to planetary influence primarily from our more distant and much more massive neighbor Jupiter. Jupiter's influence takes two main forms:

1. Its gravitational pull is continuously warping our elliptical orbit around the Sun.
2. It is similarly modifying the orbits of the asteroids (which probably number more than 30,000) so that we constitute the moving target in a celestial shooting gallery in which the bullets may be a mile in diameter.

Earth is a sphere whirling fast enough so that it bulges at the equator, flattening slightly at its poles. Even combined with the polar flattening, Earth's greatest heights and depths give no more than one-third of 1 percent distortion. That is, the Earth is out-of-round by no more than one part in three hundred; not too good for a ball bearing, but much better than many of us expected after hearing about polar flattening for years in high school.

The mean altitude of all land (including ice) is 2,887 feet (880 meters) above sea level; if the ice is not counted, it comes to 2,379 feet (725 meters). The depth of the oceans averages −12,106 feet (−3,690 meters). If the sphere were perfect, instead of having its one-third of a percent distortion, the water would cover it smoothly to a depth of 8,629 feet (2,630 meters).

Surrounding the Earth is a thin layer of air (only 4.7 percent of the Earth's radius in thickness) which has only 0.000087 percent of the Earth's mass. Human life is confined to a layer of atmosphere that is only 0.07 percent of the Earth's radius if you do not count polar flattening, 0.2 percent of the Earth's radius if you do.

Into this thick cloak of tenuous gas in which we are immersed, all living creatures and the Earth and space itself breathe.

Looking downward, we find beneath our feet a thin crust only 0.26 percent of the Earth's radius (10.6 miles, 17,000 meters) in thickness. Since the crust has a density of only 2.8 times the density of water, it bobbles around on the surface over the viscous mantle which has a density 4.5 times greater than the density of water. This viscous liquid (or molten solid) is 45 percent of the Earth's radius in thickness.

Below the mantle is a molten iron core with a density of 10.71. This comparatively small liquid ball constitutes over 31 percent of the Earth's entire mass. There is also evidence of a solid ball of some sort in the very center of the whole works.

Energy from the Sun pours onto the Earth's surface at such a rate that—despite losses from reradiation—the surface has averaged through time a temperature of 49.6°F (11°C). Records show that since man has lived in what is now the United States (35,000 years?), the Earth has been 7° colder and 2° warmer than our present average of 47.6°F. The temperature is not always this variable. There was a period of 250,000 years, 200 million years ago, which varied only about 2°.

We have casually mentioned here the idea that chunks of the crust bob like so many corks on the surface of the mantle, traveling around a lot and smashing into each other. This refers, of course, to the widely discussed theory of "floating continents" put forward chiefly by Alfred Wegener in the early part of this century. As late as 1952 college geology professors were explaining with great firmness that Wegener's theories were interesting, but had been completely discredited. The 1968 edition of the *Encyclopaedia Britannica* points out that work in the early sixties had renewed interest in Wegener. In the mid-1970s we take essentially for granted the validity of Wegener's basic ideas as they are now recast under the elegant name of "plate tectonics."

This work is so voluminously documented that it's hardly worthwhile to review the whole rationale here, beyond a statement of the basic notions. The idea is that big chunks of the crust (e.g., North America, South America, Africa) are moving in various directions fairly rapidly. The evidence is wholly convincing that Africa and South America were once joined. Africa's western indentation and South America's eastern bulge look suspiciously as if they are matching parts of a jigsaw puzzle, even to the casual viewer. Some good work has demonstrated that the geological structures on the two continents match, mile after mile. Similar matches are well documented all over the world and researchers have traced the plate movements backward through the millennia to show us plots of the organization of the world in times past. Interesting stuff.

An important question in all this is: "Where does the energy come from to drive all this action?" That is not a minor matter. It takes a whole lot of energy to force these plates to buckle and ride up on each other where they clash together.

One can imagine that the energy is somehow left over from the time when the Earth was formed, that we are still very gradually using up all that original zip. There are a number of things wrong with that idea.

One problem is especially puzzling: the rate of movement changes. We can measure the flow of the Earth's crust as it wells up out of the mid-Atlantic, for example, and we can tell how fast the material is moving away from the crack from which it issues. At times it moves fairly rapidly, then runs down as if the driving energy were being depleted; just what you'd expect. The movement seems to have stopped completely at various times, but then jumped up again very suddenly from nothing to perhaps ten centimeters a year, about four inches. That's fairly racing along, a mile every 15,840 years. It's easy to see why the system runs down. It is not obvious why it should speed up again, and not gradually, but in a step function. It is equally disturbing that the floating plates move first in one direction, then another. For example, the southern Appalachians seem to have been caused by Africa's running into America. Then the plates floated apart, with Africa donating the Carolinas, Florida, Georgia, etc., to North America. The northern Appalachians—also a buckling due to the collision—floated away to form parts of Greenland, Scotland, and Norway.

A number of reasons have been suggested for the varying rates, including radioactive heating in the Earth's core which causes convection currents in the mantle. While the energy expended in a single home-run baseball hit is enough to move the floating continent of North America ten centimeters in a year, it is another matter to lift North America or crush it. You can make a massive floating log move on the surface of a pond easily. It is more difficult to lift the log. While radioactive energy could move the floating continents, it is not clear that radioactive energy is sufficient to lift the continents.

To go back to the infrared beacon study—we had occasion at the time to do something that people seldom do. We obtained a complete set of the World Aeronautical Charts, the maps that aviators use for navigation. These are handsome, easy-to-read maps that are published in various scales: one million to one, half a million to one, and a quarter of a million to one.

Our purpose was to identify isolated sources of differential radiation. (That is, either active sources like volcanoes and fields of hot springs, or passive sources like islands, which would reflect energy differentially from the surrounding water.)

We made up some bull's-eye measuring templates, pieces of clear plastic on which were drawn circles at appropriate distances from a center point. We could drop the center of the template on an object of interest and determine at a glance how far that was from any other object of interest.

Working with maps this way is like working with a dictionary. You get so interested in the things you stumble across in your search for a particular word or place that you almost forget what you were looking for originally. You tend to accumulate a lot of fascinating and usually useless information. In this case, we discovered that our bull's-eyes of various sizes beautifully matched the shape of many structures on the maps, often structures that were not identified. We began to suspect that many of the things were not identified simply because no bored researcher before us had ever idly dropped his bull's-eye on the map in those places.

The simple fact is that there are a lot of round things on the Earth. We could determine readily that lots of these things were volcanoes, some from the ancient past. The Valle Grande of the Jemez Mountains in northern New Mexico, for example, seems to be the fourteen-mile-diameter caldera of an extraordinarily large ancient volcano. (Dead? Well, there are famous hot springs in the Jemez, and Los Alamos Laboratories are busily searching for geothermal power sources there in the mid-1970s.)

Some structures, like atolls, just happen to be round, but other things can't be accounted for that way.

Meteorite hits come to mind as the other likely source of round structures, and Barringer Crater near Winslow, Arizona, is the famous example. We began to search for other documented meteorite craters and turned up a number of them, such as Lake Bosumtwi in West Africa and Chubb Crater in Canada.

One begins to wonder after a while why the Moon is so chewed up with meteorite (or, preferably, *bolide*) strikes and the Earth is not. In fact, Earth *is* scarred by such strikes, but our atmosphere protects the surface by burning up the smaller incoming objects. Only the big ones make it to the surface and the craters that they have created tend to be masked by erosion and vegetation. The biggest ones are so big that they are difficult to detect from the ground.

The Rieskessel crater in Germany has a floor roughly fourteen miles in diameter surrounded by a thirteen-hundred-foot wall. The crater is very old, weathered, and overgrown, but it is visible under the overlying roads and human structures. During the 1940s the crater became visible to people who had occasion to fly over it in B-29s while on other business. Sixteen-inch chunks of material from this crater have been found forty kilometers away from the site. Something exciting obviously occurred there at one time and the conviction is growing that the event involved the fall of a big, heavy thing onto the Earth.

Rieskessel is but one of an increasing number of such structures identified on the Earth.

Let us recall the thirty thousand asteroids of significant size for which the Earth is a moving target. This notion of the celestial shooting gallery is dramatized by a passage from Baldwin's *The Face of the Moon*:[1]

> At first, the meteorite would plunge into the Earth moving faster than the shock waves and pushing ahead of it an ever increasing

1. Ralph B. Baldwin, *The Face of the Moon*. Chicago: University of Chicago Press, 1949.

plug of compressed rock and probably a similar plug of compressed air. When the speed of the meteorite becomes less than that of the elastic waves, the vast amount of compression produced finds a shoulder against which to push, and the mass is soon stopped.

. . . With the stoppage of motion the meteorite is sitting on top of a tremendously compressed, tremendously hot plug of matter. Naturally, an explosion of the utmost violence follows. . . .

Violent, indeed!

A one-mile-diameter asteroid, Hermes, shot past the Earth on October 28, 1937, missing us by only seventy-five Earth diameters (600,000 miles) and passing through the Earth's orbit. This space-borne mountain would have blown out over 4,000 cubic miles from a crater 80 miles across and perhaps 25 miles deep. If it had hit an ocean, it would have made a tidal wave four miles high and it would have evaporated 3,800 cubic miles of water—enough for a 1.25-inch rain over the entire Earth.

Hitting at a skewed angle, its colossal earthquake-like waves would have vibrated the Earth like a gong. The P-waves (pressure waves) would have spread out like earthquake waves, while the S-waves (shear waves) would have been absorbed by the liquid core. The turbulence would have set up a fluid flow in the core of more than a hundred miles an hour and this enormous rotating iron armature would have altered the Earth's magnetic field.

A fireball would have blossomed from thirty to fifty miles high; and there, meeting no more atmospheric resistance, it would have thrust laterally, creating a fire funnel perhaps a hundred miles in diameter at its top. A part of the 4,000 cubic miles of crustal materials blown out would have been thrust laterally to spew out and reenter the atmosphere up to a thousand miles away as tektites. A part would have been reduced to dust and put into orbit, cutting off half or more of the Earth's sunlight for years. A part would have been shot into space and would have constituted showers of stony meteorites for millennia to come.

Watson calculates that the ". . . Earth probably goes at least 100,-000 years between collisions with them . . ."[2] (asteroids). One may guess somewhat nervously from this that the Earth probably does not go very much *longer* than 100,000 years on the average without colliding with an asteroid.

2. F. G. Watson, *Between the Planets*. New York: Doubleday, 1962.

The symptoms of a great bolide strike of this sort are:
1. A switch in the Earth's magnetic field.

The switch in the magnetic field is very significant to geologists; as we have been told over and over in the popular literature, the Earth is a great magnet and its north and south poles are equivalent to the positive and negative poles on any bar magnet. Opposite poles attract. Like poles repel each other. If a positive pole and a negative pole are in contact and you change the polarity of one pole so that they are alike, the poles will jump apart. One can imagine that a sudden switch of polarities in the Earth would be an exotic event, forcing reorientation of many things on and beneath the surface. In fact, when molten rocks freeze into solid state, their polarity is fixed in the orientation of the Earth's magnetic field at the time freezing occurs. Even if a change in the Earth's polarity occurs later, the set polarity in the surface rock stays as it was.

When material is squeezed up out of the mantle through cracks like that in the mid-Atlantic, that material freezes into solid rock. When it freezes, it takes on the polarity of the earth's magnetic field at the time. If a reversal of magnetic polarity occurs in the Earth, there is a discontinuity in the polarity of the rock that has squeezed out of the crack and spread out under the ocean. Indeed, as we travel away from such a crack, measuring the polarity of the rock over which we are moving, we find that the polarity switches back and forth.

The pattern of this switching is readily plotted and it is just this sort of pattern that geologists use to establish the matching of the geological structures of West Africa with those of eastern South America. Like some other oversimplified explanations here, this one is upsetting to specialists. Still, it will serve us laymen.

2. A hard ice age.

A hard ice age may come on very suddenly. As we have indicated, the vast amount of dust suddenly thrown up by a big bolide strike might block over half of the warming sunlight from the earth for many years. We'll discuss the cooling effect in more detail later, but the simplest aspect of the mechanism is obvious: with no lovely sunshine, things get chilly.

People have long been fascinated by the fact that quite a number of mammoths have been found frozen, left over from the last ice age. Some of the creatures were quick-frozen very suddenly—so suddenly that at least one was discovered with fresh-frozen buttercups in his mouth, unchewed. It isn't easy to freeze an old-fashioned elephant whole, especially so quickly that he can't chew what he has just put in his mouth.

It has been calculated that if a mammoth were grazing in a warm field of blossoming buttercups, the temperature would have to drop to 150° below zero Fahrenheit within a period of two seconds in order to freeze the poor fellow on the spot with his mouthful of fresh blossoms.

There is a way in which that could occur. If a bolide were to strike, the explosion would heave a great mass of material up outside the atmosphere and a large amount of it would come sailing back in. A mass of dust and rock falling through the upper atmosphere would carry with it a great whoosh of cold air from the very highest altitudes. This air would be extraordinarily cold, cold enough to chill our mammoth. Similarly, if a bolide struck the Moon, the same effect would be produced on the Earth by the inrushing mass of particles.

Of course, there would be considerable turbulence as this "cold front" moved through an area. There is real evidence of this, partly in the fact that one finds more pieces of mammoths than whole mammoths. It appears that some great force literally tore many of the creatures to pieces, along with trees and other plants.

As we shall discuss later, an ice age tends to be self-perpetuating, and with a good start from a big bolide strike, such a cold period might last for a long time; this seems to have occurred 700,000 and 3,000,000 years ago.

3. A great field of tektites and microtektites.

Tektites are interesting objects, common enough so that people use them as pendants on key rings and for other decorative purposes. A tektite is a piece of rock that has been blown out of the Earth's atmosphere and has dropped back through it. The material travels so rapidly that the friction with the atmosphere during the return literally melts its forward side. This is the same sort of heating and material ablation of which we were made much aware in the early stages of the space program when we were working on reentry vehicles that would not burn up as they fell toward a target. The manned space vehicles must be carefully oriented in reentry so that their heat shields lead. The shields heat up and ablate, literally melting and even evaporating so that the hot material streams away from the surface of the shield, carrying the heat with it. Tektites characteristically show a melted, ablated face and may even be teardropped.

The beginning of the Pleistocene ice age was marked by the appearance of a wide field of tektites and microtektites in the southwest Pacific, a large drop in temperature, and a switch in the Earth's magnetic field.

A big bolide strike may throw tektites great distances, hundreds or even thousands of miles. Tektites found in Thailand appear to have originated in a bolide strike in Antarctica.

The pattern in which the tektites are discovered often points to the site of the bolide strike. Just as water splashes out in a circular pattern when you toss a rock into it, the Earth splashes out in a circular pattern when you toss an asteroid into it. By plotting the arc of the tektites, one can plot the radius of the circle and trace the things back to the apparent source of the hit. By comparing the content and structure of the tektites and samples of material from the suspected strike site, one can pretty well determine where the big object fell.

4. A giant scar, sometimes with great masses of residual material.

As we have indicated, the big bolides tend to leave big holes. Craters have been identified which seem to be a good seventy-five miles across—nominal for an average asteroid hit. A real shocker is the proposition that Hudson Bay may have in its southeastern edge a crater 275 miles in diameter. The matter is hotly debated, but throw your bull's-eye over the map and take a look. The islands that appear in a partial ring concentric with the edge of the bay are in just the place they should be if they were produced by the action of a bolide strike. Typically, the great outer rim of the crater is forced up very high by the original explosion. When things settle down, the weight of the material in that rim exerts a lot of force on the material beneath it. Since the rim is circular, its force tends to be focused inside the crater.

Craters on the Moon are often found to have central peaks that are apparently forced up by the focused downward pressure of the massive rim. If the crater is very large, the focus is not at the center of the crater, but along the perimeter of a concentric inner circle whose location can be calculated and predicted. Hudson Bay would be a fairly large crater even by the Moon's standards, and the peaks would properly be raised where those islands are.

As for masses of residual material, it appears that the International

Nickel Company is mining the remains of an old asteroid at Sudbury, Ontario. If you have a nickel in your pocket, it may well be partly from an asteroid that struck Ontario a half-billion years ago.

If the strike is great enough, it initiates a new geological age, if smaller, it produces a geological period because of its dust and colossal energy input into the Earth.

It is our great fortune that the asteroid that hit the Kamchatka Peninsula at 10:35 P.M. on February 12, 1947, was only tens of tons, rather than the average one-mile-diameter asteroid. It laid waste a herd of reindeer and scared the spit out of everybody for fifty miles around, but it did not initiate a new geologic period.

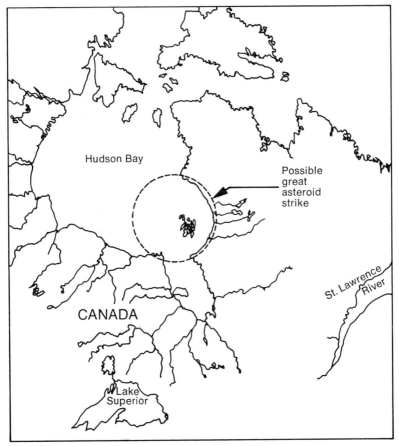

Fig. 2.1.

It has been calculated that a larger-than-average, two-mile asteroid hitting in eastern Kentucky would explode with enough energy to knock down every man-made structure east of the Mississippi. In 1968 the asteroid Icarus had a pass at us. It didn't come very close, but we couldn't be certain that it would not hit us more than a few days in advance.

We might succeed in destroying or diverting incoming asteroids, but it is more difficult to do anything practical about incoming clouds of dust, as opposed to chunks of rock or metal. Evidently some 20 percent of material that is blown off the Moon by bolide strikes may reach the Earth. The Moon's escape velocity is much lower than that of Earth, and a lot of what gets blasted out by a strike escapes the Moon. Our situation would be compromised if the Moon received a good, stout blow of the sort that made craters we can see with the naked eye from the Earth. A cloud of dust and cold air might move in on us distressingly briskly. We should have captured ten cubic miles of dust from the explosion that created the Moon's rather new crater Kepler, for instance. Note that the great volcano Krakatoa put up only one cubic mile of dust, but decreased sunlight reaching the Earth by 25 percent.

The turbulence produced by the S-waves in the liquid core after a big bolide strike would cause the liquid to rub its container, the mantle. This would produce creeping movement of the container walls. Stirrings and upwellings by the mantle would, in turn, produce the movement known as "continental drift" in the scum that is the crust of the Earth, only 0.26 percent of the Earth's radius in thickness.

A variety of crustal consequences ensue from the upwelling of the mantle:

1. New crust is formed which flows away in both directions from the upwelling zone (for example, that mid-Atlantic ridge we have referred to).
2. The new crust makes what amounts to a magnetic tape recording of the Earth's magnetic orientation as it changes through time.
3. The additional crust moving to an already covered surface takes its place by:
   a. Folding, either gross or miniature.
   b. Faulting with wedge-shaped vertically displaced blocks.

c. Great underthrusts, with consequent flotation of the thickened sectors as mountain ranges, and with downward notching to produce trenches at the junction of the override and the underthrust.

d. Lateral movement of the floating crust with buckling at the collision edge of two plates (e.g., the buckling we know as the Himalaya Mountains is an effect produced by the vigorous shoving of the tectonic plate known as India against the tectonic plate known as Asia).

4. Wheeling of blocks or plates with respect to each other, which sets up a lateral faulting line at the collision edge (e.g., the San Andreas fault in the western United States in the place where the counterclockwise wheeling Pacific plate is rubbing against North America).

Step changes in the behavior of the crust would be expected when very large bolide strikes occur, altering the turbulence in the Earth's core. The magnetic records do show such incidents as major polarity inversions in the Earth's magnetic field in the manner we have discussed.

Of course, the lateral thrust of a plate or of two plates in collision would increase plate thickness so that it would float up and have a higher average altitude. Such action has changed the mountain states from a shallow ocean in relatively recent geological times into the Rocky Mountains.

Upthrust wedges or upended plate fragments bring deep deposits to the surface. Erosion of such newly exposed segments redistributes trace elements and greatly enriches the land. The newly eroded soil of the valleys of California makes them among the richest on Earth, whereas the geologically nearly dead deserts of Australia are infertile more because of the absence of trace minerals that have been leached away than because of lack of rain. Living things require trace elements.

Since the crust is a true solid, it moves not in a fluid flow, but in a rasping, jumpy, crunchy fashion. Each of these movements results in an earthquake.

The great underthrusts not only produce a series of motion shocks, but they also melt or trap pockets of fluid that spew or pop out as volcanic eruptions.

As we begin now to talk about the release of energy from the Earth, we'll use "scientific notation" to express numbers and we'll speak of energy in terms of ergs. Most of us speak this scientific tongue only haltingly and not many of us have any feel at all for what an erg is. Some translation and explanation is in order.

An erg isn't much. It's roughly the amount of energy needed to lift one gram one centimeter. (Roughly, because the weight varies a bit with latitude.) A gram isn't much, either, and a centimeter is only about two-fifths of an inch. A thousand grams is about equal to two and a half pounds. Divide two and a half pounds of butter, for example, into a thousand equal parts. If you lift one of those parts two fifths of an inch, you'll be expending one erg of energy, not counting the energy you spend laughing at the process.

For Americans, this still isn't much help. Consider this: If you drop one pound of butter one foot, it will strike the table top with a force equal to 13,560,000 ergs. See? An erg isn't much.

However, we'll be talking about very large numbers of ergs. Scientific notation is a shorthand for writing and manipulating very large numbers. For example, the 13,560,000 number above is written 1.356×10^7 in scientific notation.

The number is multiplied by ten every time we add a 1 to the exponent above it. The exponent tells us how many zeros come after the 1 in the 10. 10^2 is 100—two zeros come after the 1. 10^{25} would be written out as 10,000,000,000,000,000,000,000,000. Only when you get to comparatively large numbers does scientific notation pay off, but we'll deal with large numbers and it's worth it. Though an erg is trifling, 10^{25} ergs is a whole lot of energy.

Table 2.1 lists a number of events or processes, from small to large, indicating the number of ergs expended in each event. We'll refer to some of these events from time to time in the course of the book, and the table will provide some sense of the magnitude of the event in relation to other things.

Consider the estimate that earthquakes release 10^{26} to 10^{27} ergs per year on the average at the present.

A single bolide only 900 feet in diameter, of typical nickel-iron composition, colliding at the nominal speed of ten miles per second, would expend 10^{26} ergs on the Earth. This would equal 240 million tons of T.N.T. or ten Krakatoa volcanoes.

A nominal (one-mile diameter) asteroid strike would expend 1.5 million million tons of T.N.T. in explosive energy. This would be

Table 2.1.	
Energy in Representative Events	*Ergs per Event*
One gram dropped two-fifths of an inch	one erg
Pound of butter (or anything) dropped one foot to a tabletop (one foot-pound)	1.356×10^7
Moonquakes	10^9 to 10^{12}
Home-run hit of a baseball	10^{10}
Earthquake magnitude 0	2.5×10^{11}
Earthquake magnitude 1	7.9×10^{12}
Earthquake magnitude 2	2.5×10^{14}
Earthquake magnitude 3	7.9×10^{15}
Lightning bolt	10^{16}
One ton T.N.T.	4.16×10^{16}
Earthquake magnitude 4	2.5×10^{17}
Earthquake magnitude 5	7.9×10^{18}
Earthquake magnitude 5½	4.4×10^{19}
Earthquake magnitude 6	2.5×10^{20}
Earthquake magnitude 7	7.9×10^{21}
Siberian meteorite (1908)	1.8×10^{23}
Earthquake magnitude 8	2.5×10^{23}
Fusion bomb (U.S.S.R.) (60Mt, probably an accident)	2.5×10^{24}
Earthquake magnitude 8.9 (among largest recorded)	5.6×10^{24}
Earthquake energy, average annual	9.0×10^{24}
Karakatoa volcano (1883)	10^{25}
Mean path length of annual polar wobble 1 arc sec/year represents annual energy change of	3.0×10^{25}
Tambora volcano (1815) (25,000 megaton bombs)	8.4×10^{27}
"Nominal" (i.e., 1-mile diameter) asteroid collision	6.0×10^{28}
Explosive energy of largest known probable asteroid collision crater on Earth (S.E. edge of Hudson's Bay)	10^{30}
Energy required to raise Andes Mountains	10^{31}
Asteroid collision required to form Mare Imbrium on Moon	8.0×10^{31}
Erosion per geological period (~ 60 mil. yrs.)	10^{32}
Total earthquake energy per geological period with ~ 10-milion-year half-life of forces causing the earthquakes	1.8×10^{32}
Large solar flare	3.0×10^{32}
Total earthquake energy per 60-million-year geological period at present rate	5.4×10^{32}
Energy absorbed by liquid core at 10% efficiency from following asteroid pulse	5.4×10^{32}
Input pulse by a 45-mile diameter asteroid (20 mi. per sec.) colliding with Earth	5.4×10^{33}

about 6×10^{28} ergs. A great (10-mile diameter) asteroid strike would be a thousand times greater—6×10^{31} ergs ($6 \times 10^{28+3}$).

If only 1 to 10 percent of the annual average budget of energy expended on earthquakes were concentrated in one area as an uplift, that would provide 10^{25} ergs per year for raising mountains in that area. That small percentage of the available energy could raise the equivalent of the Andes Mountains in just a million years—10^{31} ergs (10^{25+6}).

If all of this earthquake and mountain-building activity comes from turbulence in the core, the turbulence will clearly run down as its energy is transferred from the liquid to the mantle. Allowing for a ten-million year half-period of "running down" (we may think in terms of a set period in which a system expends half of its energy this way, just as we may think of the "half-life" during which a radioactive material expends half of its radioactive energy), the amount of energy expended would equal 2×10^{32} ergs during an average 60-million-year geologic period. This amount of energy could be delivered by a *single* hit from one extra-large 30-mile-diameter asteroid.

The geophysical energy expended by the Earth in a short one-million-year period is about twenty times the amount of energy required to raise the Andes in the same time.

The Earth has had fourteen mountain-building periods that we can identify. The energy for this could have been supplied by fourteen great hits.

An interesting check on these calculations is a calculation of the energy expended in erosion. Erosion is a matter of having material that has been lifted "fall" back down again to the level from which it was originally lifted. The stuff expends just as much energy when it comes down as it absorbed when it was being lifted up.

One can calculate reasonably that, by the end of a 60-million-year geologic period during which mountains have been lifted way up and then eroded all the way down again, the energy involved in the erosion alone is about 10^{32} ergs—half of the energy we calculated as being transferred from the core to the mantle in that time. That's not a bad balance, considering losses from radiation, heat, etc., which could account for the other 10^{32} ergs. Any time you come within an order of magnitude (a factor of 10, or just 1 in the exponent) you feel as if you have a good ballpark estimate of the real values. Having things as close as a factor of two, as we have here, makes the figures seem very close indeed.

All of the energy we have talked about here could be supplied by known physical mechanisms. Nothing outlandish is required. A modest 1.4-cubic-mile asteroid would supply enough disruptive force to terminate geological periods and, presumably, initiate succeeding ages. It is unlikely that blows of the sort we have discussed would involve reorientation of the Earth's axis, a possibility that always occurs to people who think about big, heavy things smashing into the Earth. We ride on a quite placid gyroscopic flywheel.

We needn't rely on occasional events like blows from exceptionally large asteroids every 60 million years. We can rely on a reasonably steady rain of more modest asteroid strikes every 150,000 years or so to provide all the energy we need to keep old Earth splashing and grinding.

It can't be very much fun to participate in a big strike, and one suspects that certain peculiar discontinuities in the records of living things are associated with these events. On the other hand, man was around when the last really big one hit us, 700,000 years ago—as were dogs, cats, and many other familiar animals.

The strikes plow the Earth for us on a grand scale, turning over the crust so that the rich materials underneath are brought repeatedly to the surface to nourish living things. This may be nerve-racking, but it lends spice to life.

If man were to learn to stop these bullets by intercepting them in space with ICBMs, and were to prevent them ever from coming in, the Earth might well run down.

In 20 million years, erosion would reduce North America to low-lying, rolling terrain. The earthquakes and volcanos would have died out. Earth would be hot, because no steady, protective shield of dust and sulfur dioxide would be fired into the stratosphere to reflect sunlight. The steaming swamps and fern forests of Kansas and Iowa would hold evolved life forms in a steady state. These would compete with the decadent descendants of man.[3]

Man himself would not be stimulated by droughts, extremes of climate, or other environmental change. He would sink into the long slumber that the world went through once in the past, when the giant lizards and things that slithered reflected the general state of the Earth.

3. Lee Parman suggests in a note to us: "This malady of decay would contain its own cure, however. Under a steady, equator-like environment, man would trade understanding for taboos, science for custom, creativity for compliance, and freedom for security. He would dwell for so long in an intellectual twilight that he would lose the knowledge, and the bolides would begin to land again."

3

We have been talking about astronomical phenomena that have striking effects on Earth. There are other astronomical matters that have an important, if not quite so dramatic, effect as bolides. Tidal forces on the Earth are governed by the movements of astronomical bodies.

People have paid a lot of attention to tides over the centuries because of the ancient insistence on going out on the ocean in boats. The waters rise and fall at the shores as the tides flow in and out. It makes a lot of difference to your health and to your schedule whether the water is coming or going as you try to maneuver a boat close to the shore. Mariners have gone to great pains to figure out what the tides will be like at particular times so that they can do something sensible with their boats.

Oddly, our intellectual heritage in the Western world did not feature major works on the tides until the time of Newton. The "classic" literature of the Greeks and the Romans paid little attention to the subject, because the cultures in which that literature was produced centered on the Mediterranean Sea—a body of water in which tides are insignificant. Odysseus sailed the wine-dark sea without realizing that he wouldn't have qualified as a very good sailor in a large part of the world. When Shakespeare, an Englishman who watched the tide rush in and out of the Thames every day, gave Cassius the speech

"There is a tide in the affairs of men which, taken at the flood, leads on to fortune . . ." he was using poetic imagery with great significance to Englishmen. However, the real Cassius wouldn't have known what he was talking about. Caesar could have told Cassius something about tides. Caesar had experience with them, which took him by surprise when he went off to conquer England some time before Shakespeare. The surprise was big enough so that he commented on the matter in his writings. Similarly, when Alexander the Great made it all the way down the Mediterranean area to the mouth of the Indus River in the Indian Ocean, he was startled by the great variations in sea level.

The height of tides varies around the world from practically nothing to fifty feet or more. In some areas, if you tie your dinghy to the pier with a short rope at high tide, you'll find the thing dangling like a fish on a hook as the tide goes out. It pays to think ahead—and people *do* think ahead in this matter. Enormous efforts are expended to figure out what tides will be like a year or more ahead of time at ports all over the world.

It is very easy to hear a lot more about tides than you really want to unless you are an enthusiast on the subject. The complex wave motions of the waters fascinate a layman only to a certain point. Most texts and encyclopedia discussions of the tides explain the basics in plain language and then break into completely unintelligible but impressive formulas for harmonic analysis. Our interest in the movement of the waters is limited here and we'll refrain from the poetry about harmonic analysis, but a number of points are relevant and worth mention before we move out into space again.

Extremely precise measurements of ocean tides are hampered because there's no "solid" place to stand while you measure the change in water level. The fact is that the land responds to tidal forces just as the water does. The granite on which you confidently stand to make your measurements may be rising up and down in tides as great as two feet.

Not only is the surface of the ocean made irregular by the waves that ride on the tide and by the wind and by variations in atmospheric pressure, but the *bottom* of the sea, along with the Himalayas and the Alps, is sloshing up and down. This makes it very difficult to do even indirect measurements, which depend on changes in gravity owing to shifts in mass caused by tides. With everything moving, you can't quite tell what's going on. It isn't that Earth is just a great ball of Jell-O, quivering nervously at every touch; it has been calculated that Earth as a whole is about as elastic as a ball of steel. However, the forces are very great and the ball is very large, so that even minute percentage changes in its dimensions seem gross to us who have no other place to stand.

Great forces? How great—on a personal basis, that is?

Alfred Defant suggests that a two-hundred-pound man standing dead center on the point of highest tidal pull would be reduced in weight by about ten milligrams. As Defant helpfully explains it, that's about equal to one drop of sweat.[1]

Like most theoretical models, the splendid mathematical representations of tides have only limited practical application. The real Earth is so much more complicated than anything that can be handled in a formula that the predictions of real oceanic tides are a mixture of complex math and practical observations loaded with fudge factors.

The theoretical model is even less practical when it comes to tides in the land, where the structures are far more complex even than the oceans. It is extremely difficult to do useful calculations on something with an essentially infinite number of variables.

It is with tidal forces in the land that we are most concerned, however, and we'll settle for dealing with the matter in a general way.

Let's leave the ocean bottoms and move away from the Earth for a longer view of things.

Apart from a bow to Jupiter, we can limit our discussion to the three bodies Earth, Sun, and Moon. Other planets are too far away, with too little mass to have major influence in the forces that interest us.

Earth orbits around the Sun in an ellipse at an average distance of 92.9 million miles. Since the orbit is not perfectly circular, with the Sun right at the center, we move a bit closer to and a bit farther from the sun at different times of the year. At the closest point, called perihelion, we are about 1.7 percent closer to the Sun than the average, and at aphelion, we are 1.7 percent farther away than the average

1. Alfred Defant, *Ebb and Flow*. Ann Arbor: University of Michigan Press, 1958.

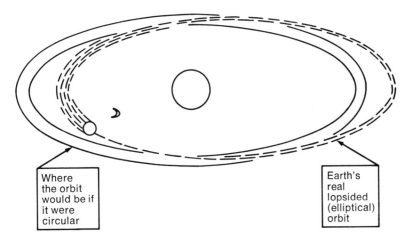

Where the orbit would be if it were circular

Earth's real lopsided (elliptical) orbit

Closest approach to Sun by Earth is called perihelion.

Farthest retreat from Sun by Earth is called aphelion.

Fig. 3.1.

distance. This gives us a variation in distance over the year of about three million miles. Perihelion falls in early January and aphelion in early July.

Three million miles out of 93 million is a significant amount. The force of gravity—the pull between bodies like the Sun and the Earth —changes inversely with the square of the distance of the bodies. That is, if we say that the force of gravity has a value of 100 when the two bodies are 1 million miles apart, then the value for gravity is only 25 when the bodies are moved to 2 million miles apart, and only 11.111 (one-ninth of a hundred) when they are moved to 3 million miles apart. This means that a small change in the distance produces a comparatively large change in the force operating between the two objects.

The Sun is a very massive body (indeed, it contains better than 99 percent of the mass of the entire solar system), and even though it is so far away that it takes light over eight minutes to reach Earth from the Sun, the Sun's gravity does affect events on the Earth.

The Moon is much smaller, but it's much closer, only 238,857 miles away on the average. Light travels from Moon to Earth in about 1.3 seconds. At its closest point, perigee, the Moon is 221,463 miles away, and at its farthest point, apogee, 252,710 miles away. That makes for a gross difference of 31,247 miles in the distance of the Earth from the Moon during each lunar orbit. (This period from perigee to perigee is called the "anomalistic month.") The total variation, then, is about 13 percent of the average.

The tide-generating force of the Moon is about twice as great as that of the larger but more distant Sun, and the variations in the Moon force are very much greater.

The strength of the tide-generating forces on the Earth at any given time clearly depend on three basic factors: (a) the distance of the Sun from the Earth, (b) the distance of the Moon from the Earth, and (c) the alignment of the three bodies.

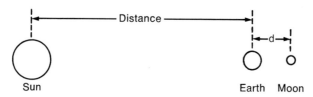

Fig. 3.2.

If all three are perfectly lined up, the forces of Sun and Moon add up linearly to a very large tidal force. It doesn't matter whether the Moon is between Sun and Earth or beyond the Earth. The forces add up the same.

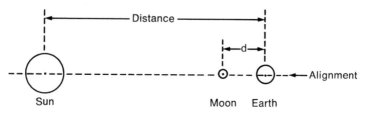

Fig. 3.3.

It does matter how far apart the bodies are.
Earth may be at perihelion and Moon at perigee.

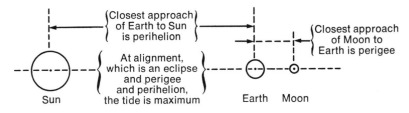

Fig. 3.4.

Earth may be at perihelion and Moon at apogee.

Fig. 3.5.

Earth may be at aphelion and Moon at apogee.

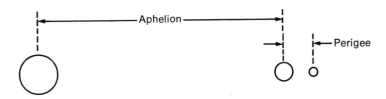

Fig. 3.6.

And so on. The permutations are obvious.

Notice that we have an eclipse in each of these cases illustrated because either the shadow of the Earth falls on the Moon or the shadow of the Moon falls on the Earth, depending on which is in the middle.

We have been describing here the configurations with the highest tidal effect on Earth. The really potent arrangement is that first one —perihelion, perigee, perfect alignment (eclipse). That gives us an enormous tidal surge when it occurs.

It doesn't occur often, and for folks who like balmy weather that's a good thing.

Notice that the anomalistic month we have referred to is 27.554551 mean solar days long, just about four weeks. That means that if the Moon were in the middle of the set, casting its shadow on Earth to give us an eclipse of the Sun on a particular day, it would show up on the opposite side of Earth just two weeks later. The Earth's shadow would fall on it there, giving us a nice eclipse of the Moon.

It happens that alignment is not this simple. You may have noticed that we don't actually have eclipses every two weeks. (Events in the heavens are not always noticed. When a star blew up in 1054 to form a nova that is still with us as the Crab Nebula, folks in China, Japan, and Arizona commented on it, but not in Europe.)

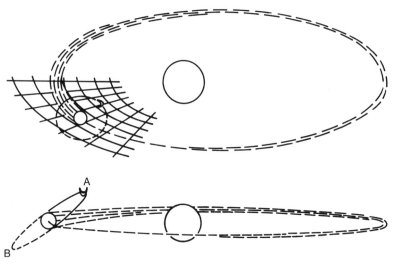

Fig. 3.7.

The orbit of the Moon around the Earth is not in the same place as the Earth's orbit around the Sun. It's cocked at an angle.

The Moon crosses the plane of the orbit of the Earth twice in one trip around the Earth, but it doesn't linger there, just passes through. The time it takes the Moon to make the full circuit from one such crossing point through the other crossing point *and back* to the first is called an eclipse month or, more colorfully, a draconic month. Its length is 27.212220 solar days. Again, the Moon passes through the Earth's orbital plane about every two weeks.

Obviously, the tidal forces of Sun and Moon don't add up linearly when they aren't in a line. The angle of misalignment becomes important along with distance. In Figure 3.7 we showed the Moon at one of its two best alignments in this trip around the Earth. The other best alignment is down at point B.

We should here identify a third type of lunar month, the synodic month. This is the ordinary lunar month, the period between full Moons, 29.530589 solar days.

The Moon is full when it is out beyond us, away from the Sun, and we can see the whole lunar hemisphere which has sunlight falling on it. The new Moon occurs when the Moon is closer to the Sun than the Earth is and we can see only the unlighted side. These times of new and full Moon are also the times of best Sun-Earth-Moon alignment. It has been observed from very ancient times that the height of the ocean tides follows a fairly regular pattern related to new and full Moons.

Another movement is important to us. Point A and Point B in Figure 3.7 are not always in the same places. That is, they don't always coincide with full Moon and new Moon. The Moon's orbit around the Earth "precesses" so that the Moon cuts through the plane of Earth's orbit at slightly different points at every cycle.

In fact, the Moon will work its way around to points A_1 and B_1 (see Figure 3.8) again every 18 years and $10\frac{1}{3}$ days, depending on the number of leap years in the period. This period is the "eclipse node cycle."

It happens that the anomalistic month (perigee to perigee) fits into this period almost an even number of times, so that when the Moon gets back to point A_1 it will be almost the same distance from Earth as it was last time.

Suppose we put points A_1 and B_1 right on the eclipse nodes in the plane of the Earth's orbit. A_1 and B_1 would then mark eclipses.

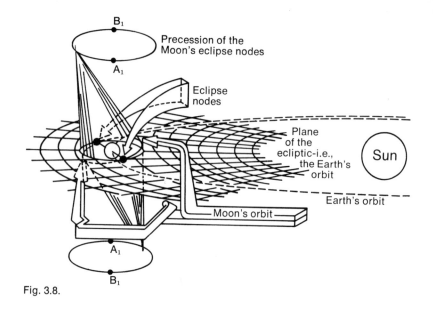

B₁

Precession of the
Moon's eclipse nodes

A₁

Eclipse
nodes

Plane
of the
ecliptic-i.e.,
the Earth's
orbit

Sun

Earth's orbit

Moon's orbit

A₁

B₁

Fig. 3.8.

Eclipses indicate good alignment.

Good alignment indicates straightforward addition of the tidal forces of Sun, Moon, and Earth.

Before we get down to explaining what this has to do with our main subject, let us mention some additional confusion factors so that nobody will conclude that anything about this is simple.

Eclipse enthusiasts talk about yet another sort of month, the sidereal month, the time it takes the Moon to make a complete circuit of the Earth with respect to the stars—as seen from a distant galaxy, for example. They speak also of the eclipse year of 346.62 days, which is the interval at which the Sun crosses the same eclipse node. (Our conventional year is 365.242194 days, during which the Sun goes from perihelion to perihelion.)

We make our bow here to Jupiter, which warps the Earth's orbit around the Sun, changing the difference between perihelion and aphelion. About twenty thousand years ago that difference was appreciably greater than it is now, making for a greater range of variation in the Sun's tidal forces acting on the Earth. Twenty thousand years from now the range will be even smaller than at present, but in 40,000 years it will have increased again and it will continue to increase through the following 60,000 years at least.

That is to say: Everything is changing with respect to everything else. All predictions of eclipses, tidal forces, and miscellaneous items of interest like the orbits of asteroids are very complex. Luckily, people have been clever in making predictions of events that seemed significant to them.

As long ago as four thousand years ago the Englishmen of the time dragged a lot of huge rocks out onto the Salisbury Plain and set up a calculator for the prediction of eclipses. We call that structure Stonehenge now, and some of us are impressed not only by the accuracy of the calculator, but by the trouble those folk went to over eclipses. They must have considered eclipses extraordinarily important and they were very anxious to know exactly when they would occur.

We tend to shrug off our ancestors' concerns as superstitious awe. It is easy to assume that the building of Stonehenge (and some 900 similar structures in the British Isles) was prompted by childlike fear of blood-red Moons in eclipse and darkness at noon. Presumably the archeologists in a few thousand years will unearth *our* most durable remains, freeway cloverleafs and porcelain toilet bowls. They will attribute our interest in these things to dim-witted superstition, probably, though we regard them as practical. The ability to anticipate the occurrence of eclipses—good tidal force alignments—may have seemed so practical to ancient man that he was willing to spend a lot of energy on the matter.

We have developed even better calculators and we are able to predict overall tidal forces for many centuries into the future, as well as plotting them far into the past.

We are concerned with perihelion and perigee as well as with the alignment indicated by eclipses, but alignment appears to be the most important factor in tidal forces. We may assume that very high tidal forces are never associated with bad alignment.

Eclipses, then, are pointers to high tidal probabilities, which depend in addition on the distances of the bodies.

It is important to remember in this game that near misses *do* count. As the bodies gradually approach the points of best alignment, the Moon makes pass after pass, closer and closer to the high point. Then it passes the point and the goodness of alignment gradually decreases in pass after pass.

As we have indicated, the movements of the bodies are not perfectly cyclic. Orbits distort and shift over time. However, the changes are not fast and over periods of some thousands of years the same patterns of alignment combine with the same patterns of separation among the bodies to produce tidal effects upon the Earth which *are* grossly cyclic.

In certain eras, then, we have many eclipses, many near misses, many good alignments coinciding with perihelion and perigee. In other eras the alignments and coincidences are few and the tidal forces are generally lower, with smaller dynamic range. The anomalistic, draconic, and sidereal months drift in and out of phase with each other, with the solar year, and with the eclipse cycle to create changing times of great and small tidal forces.

To bring this closer to home, imagine that a neighbor has prevailed upon you to look after his pets for him while he is off on a ten-day trip. You agree because you are hoping for a favor in return at some later time, and he gratefully gives you the schedule for three tasks.

1. The tank in which he breeds prize guppies must be checked *every six hours, day and night,* and any new little fish must be netted and placed in another tank before the parents eat them.

2. The cat, for reasons best known to him, must be allowed to step out in the back yard for a few minutes *every nine hours, day and night.*

3. A couple of turtles are sick and it is necessary to put a drop of medication into their tank *every ten hours*, day and night.

In addition, though you are too polite to mention it, you must take an allergy pill yourself *every four hours*, day and night, or the association with the cat will make you ill.

With tasks scheduled at four-, six-, nine-, and ten-hour intervals, you know that you can never sleep for more than four hours at a stretch, but that doesn't seem a hardship and the four-hour interval fits easily inside all of the other periods.

At midnight you take your allergy pill and step next door to go through the tasks with your neighbor once before he rushes off to catch his plane.

He is scooping some baby fish out of the guppy tank as you arrive. The cat scoots out the back door as you open it. Your neighbor shows you where the turtle medicine is and you put a drop in their tank.

Nothing to it. As you leave the house the cat runs back in through the open door. You go home and nap.

At four in the morning the alarm goes off and you take your allergy pill. At six the alarm goes again and you hurry over to net a few more baby guppies.

At nine you hurry back over to give the cat his five-minute break in the back yard.

At ten the turtles get their drop of medicine.

At noon you take another allergy pill and net some more guppies.

The next four hours are free and you get another good nap, but at four you take another pill, at six you net more guppies and let the cat out, at eight it's pill and turtle time again.

It occurs to you now that you should have thought more carefully about this schedule ahead of time and taken steps to improve it, perhaps by staggering the tasks instead of starting them all at the stroke of midnight on the first day.

Indeed, if you make a plot of your activities over the whole ten-day period, following the schedule you have already got yourself committed to, you realize that the tasks tend to pile up systematically at noon and midnight to produce peaks of frantic activity. At noon on the eighth day all four tasks occur at once, as they did that first midnight.

Worse than that, on the third day there's only one four-hour rest period in the whole twenty-four hours. You'll get very tired during that period, but by the time your neighbor gets back at the end of the ten days, you'll look fresh and relaxed because there are five four-hour rest periods for refreshing sleep during the last two days, with three of those in the last day. Unless you force yourself to stay awake to preserve the effect, you won't even look haggard and weary enough to impress him with your hard work and devotion.

If you do try to plan more carefully, you will have different but not necessarily more satisfying results. Suppose that you see your neighbor off at midnight, wait an hour, take your allergy pill at one, net some guppies at two, let the cat out at three, and medicate the turtles at four.

This does stagger the tasks and lock you into a different pattern. As the chart shows, you won't get a four-hour sleep period until the middle of the third day and you'll get none the day after that. You never have more than two tasks to do at a time, but the frequency of the tasks jumps up sharply so you are constantly running next door.

The far more complex tidal patterns affecting the Earth are pro-

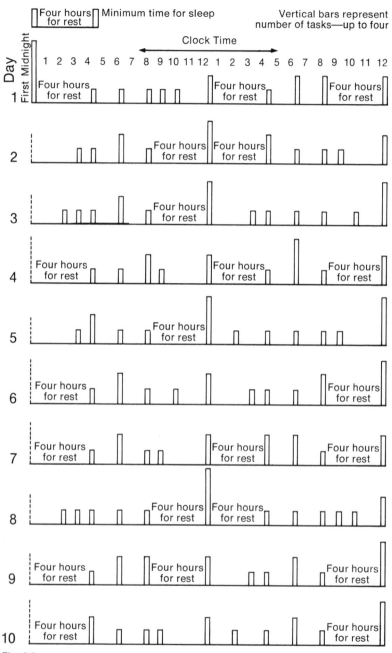

Fig. 3.9.

Alternate Plan

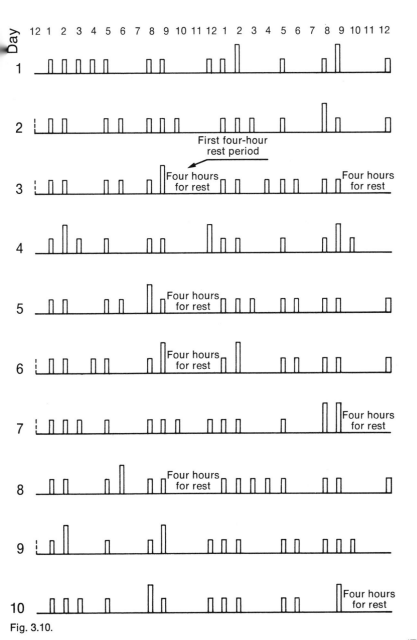

Fig. 3.10.

duced in just this same fashion. Instead of depending upon schedules that call for action every few hours, the tides depend upon factors like these:

Anomalistic month	27.554551 mean solar days
Synodic month	29.530589 mean solar days
Draconic month	27.212220 mean solar days
Perigee precession	8.85 years
Eclipse node precession	18.6 years
Conventional year	365.242194 days

These periods all work together, reinforcing or weakening the tidal forces on the Earth according to consistent, predictable patterns. At times the forces add up frequently to high peaks. Sometimes there are long spells of quietude, with little variation in the forces and consistently low levels of tidal activity.

We have observed effects that occur at intervals of 4.425 years, 8.85 years, 9.3 years, 18.6 years, 45 years, 177 years, 350 years, and 801 years. These effects seem to be strongly related to the tidal forces that have operated over the last few thousand years—and in one case over hundreds of millions of years.

It seems that these same periods will be effective for some time into the future, probably for several thousands of years, surely for some hundreds, and certainly for the few scores of years that will be of personal interest to those who are now living.

4

On February 8, 1971, we arrived in Los Angeles to make a consulting date the following day in La Habra. As we drove the weary route to Orange County, the evening radio reports mentioned an earthquake in Venezuela and a volcanic eruption in another Latin American country. Dr. Browning cheered up. "Well, tomorrow's a great day for an earthquake."

"Why?"

"For one thing there's an eclipse of the Moon."

"I wonder why we left the nice, peaceful Rio Grande Valley and came to the San Andreas fault for the occasion."

"Well, there's no telling within twelve thousand, five hundred miles where the action will occur. Enjoy it."

At the dinner with our clients that evening, we discussed the eclipse and the volcano and the earthquake and went through a brief discourse on tide forces. The discussion added a bit of spice to the meal, but a salesman of earthquake insurance wouldn't have found his market much improved by the proceedings.

As we parted, somebody asked what time the quake should be expected. "About six in the morning or six in the evening." Fine, fine. Good night.

At 6:07 the next morning the famous San Fernando Valley earthquake saved the motel switchboard operator the trouble of waking us up. Some of us who have been through a good many quakes have never learned to enjoy them.

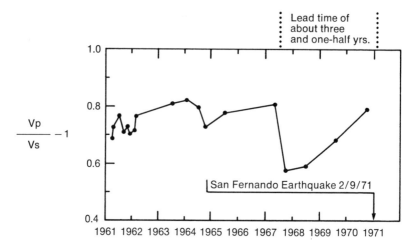

Fig. 4.1.

Vp is the P-wave velocity; Vs is the S-wave velocity.

Each data point was obtained from Vp and Vs from 19 earthquakes between 1960 and 1970 of magnitudes 2 to 4 and in line with the two seismic stations at Pasadena and Riverside.

There was a total eclipse of the Moon on February 10, 1971, the middle of the eclipse being at 7:43 UT, duration 6 hrs. 13 min.

A tidal maximum occurred on February 8,9, 1971, at 30° N. latitude.

After Whitcomb et al, 1973

By the time we got to the lab, our dinner companions of the night before were ready for us. Ed Sei was complaining that he had poured a cup of hot coffee all over himself because our warning hadn't been stern enough to make him take the matter seriously. Don Isenberg reported that the shaking had wakened his wife, who said, "What's that?"

"Aw," said Don, "it's an earthquake. Iben told me there was gonna be one, but I forgot to mention it."

While the illusion of prophecy we had created was fun all that day, it was no great fun to look out from the low motel building we selected near the L.A. airport that evening and watch the Earth's shadow creep across the Moon, to the accompaniment of many creepy little aftershocks.

Can we predict quakes, really?

Well, not exactly.

Was that pure, dumb luck?

Well, not exactly. But we can give you some information to work with and you can play your own guessing games.

We have already talked about a mechanism by which a great amount of energy may be pumped into the Earth—big bolide strikes.

Plate Tectonics

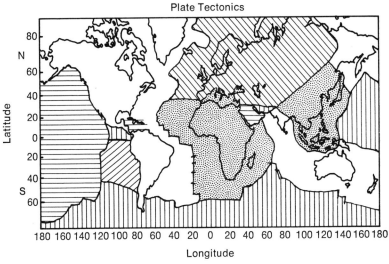

After Van Houten, Science *166*

Fig. 4.2.

We have talked about the movement of Earth's crust, not in smooth flow, but by fits and starts. These movements release the greater part of the stored energy, while the rest is dissipated as heat, etc.

1. There are crustal plates that move about on the Earth's surface ("tectonic plates," "continental drift").
2. There is upwelling of the mantle along certain lines on the Earth's surface. This produces new crust and forces the plates away from the upwelling.
3. Both earthquakes and volcanoes occur along such lines of upwelling.

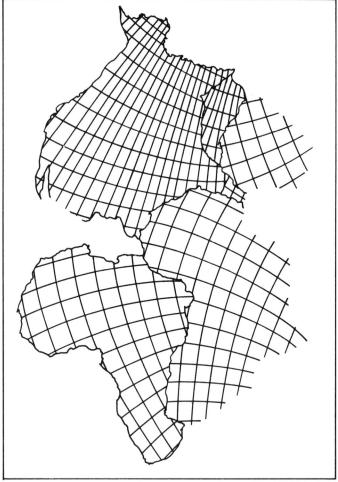

Fig. 4.3.

After McWhirter 1964

4. There are zones of crumple and zones of underthrust. Each of these is directly associated with earthquakes. Underthrust is normally associated with volcanoes some distance away through the overriding plate.
5. There is tectonic plate shearing. The shearing edge (e.g., the San Andreas fault) is the site of longitudinally active earthquakes. This is *not* normally associated with volcanoes.

Figure 4.4 shows the general distribution of earthquake and volcanic zones. One notable area, about which we hear a lot, is the Great Pacific Rim of Fire, which can be traced from Antarctica, clear up the coasts of South and North America, along Alaska and across the Bering Sea to Kamchatka, down through the Japanese islands, swinging through the East Indies, then back through the South Pacific islands, *missing Australia,* and down through New Zealand.

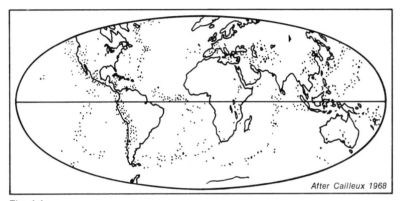

After Cailleux 1968

Fig. 4.4.

A branch of that system runs down from the tip of the Aleutians through the Hawaiian Islands. The Hawaiian Islands are a chain oriented roughly east and west with the Big Island, Hawaii, at the eastern end.

The island of Hawaii is volcanically very active, with Mauna Loa and Kilauea erupting spectacularly often. As you head west through the chain, the volcanic action diminishes and the configuration of the islands changes. Hawaii is the youngest island in the chain and the others appear to grow progressively older as you move west. The change in configuration is produced by erosion.

It appears that the islands are popping up through a weak place in the crust, now centered under Hawaii, and are then sailing away like a string of toy boats to the west. As they sail away from the hot spot, they cool and calm down until they grow quite serene.

53

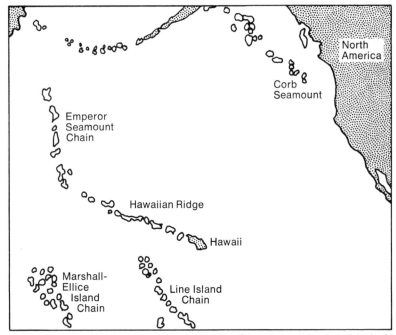

After Dalrymple et al, American Scientist *61*

Fig. 4.5.

A second area of note is the mid-Atlantic ridge, of which we have spoken. This runs from Iceland down through the Azores, the Cape Verde Islands, Ascension (of whose charms many a World War II serviceman will tell you with great bitterness if you give him half a chance), St. Helena (of which you may gain some idea when you realize that they sent Napoleon there when they decided to get rid of him once and for all), Tristan da Cunha (whose tiny population was temporarily evacuated in the early sixties, owing to volcanic excitement) and on to Antarctica.

The Great African Rift is another volcanic zone, where the Mountains of the Moon and other ranges have pushed up out of the crack to dam up the great string of African lakes. Mount Kilimanjaro, Niragongo, and Nyamlagira are part of this chain. (Great romance is associated with some peaks. Hemingway and Ruark wrote of Kilimanjaro. The great runner Kip Keino builds his endurance by

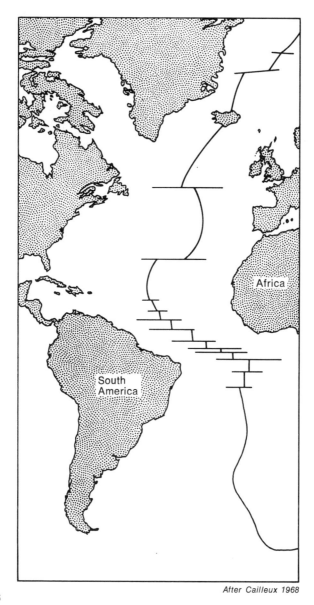

After Cailleux 1968

Fig. 4.6

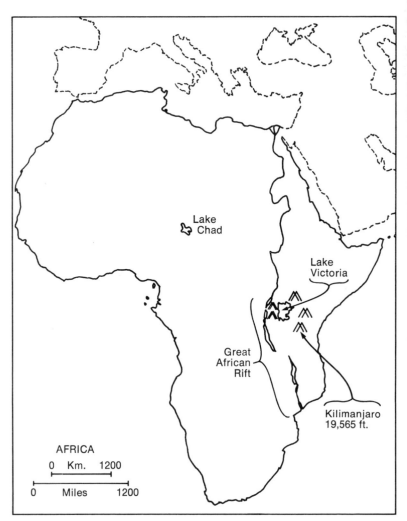

Fig. 4.7.

running on its slopes. Bobby Kennedy and his friends climbed Kilimanjaro with some fanfare and the added excitement of a casualty who died of the exertion at almost twenty thousand feet.)

Another strip of volcanoes sweeps across Italy, Greece, Turkey, and the Caucasus, where the peak we now identify as Ararat rears its volcanic head.

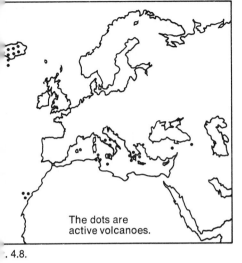

The dots are active volcanoes.

. 4.8.

The dots are active volcanoes.

Fig. 4.9.

Yet another major strip reaches down from the Arabian Sea through the Indian Ocean, including volcanic islands like Rodriguez, Reunion, and Heard. (Heard is an icy, craggy, depressing bit of property that nobody seems to like. An English party setting up a weather station there had to be rescued suddenly when the volcano let go—also in the early sixties.)

Smaller branches of these main volcanic lines occur here and there, with a scattering of odd loners. Two very interesting loops of volcanoes occur in the Caribbean at the top of South America and in the South Atlantic, at the bottom of South America. These loops have the same general configuration; it's as if a great force were extruding the Earth's crust eastward through the "weak places" that occur at the narrow isthmus of Central America and the gap between Antarctica and South America.

Volcanoes do get man's attention when they let fly. However, people become strangely accustomed to the presence of volcanoes, even when the things are giving signs of digestive upset.

Picnickers frolicked in a crater of Krakatoa just a couple of days before a major eruption, fascinated by the rumblings and growlings, but reassured by the lovely vegetation.

Many of us old-timers remember the love affair that erupted on the front pages when Roberto Rossellini took Ingrid Bergman to the Italian island of Stromboli to shoot a film. Stromboli obliged by erupting, too. How many of us remember that a gentleman of the film company wandered into a hollow full of poisonous gases while scouting locations there and died of the effects?

When a volcanic island appeared in Japanese waters with a great racket of explosion and steam, then sank from view, to rise and fall repeatedly for some weeks, a boatload of Japanese newsmen sailed out to record the event. They couldn't find the thing until it came up under them and they became part of the news, posthumously.

When Vesuvius broke tradition by erupting after uncounted centuries of quiet in A.D. 79, Pliny the Elder rushed across the bay to lend a hand. It got him. His nephew, Pliny the Younger, wrote about it eloquently and we read his descriptions with interest today. That eruption buried Pompeii.

When Mount Pelée began to grumble and produce little quakes on the Caribbean island of Martinique in 1902, some folks in the city of St.-Pierre grew nervous. Others, naturally wiser and more calm, laughed at these petty cowardices and called for order. Notable among the calm group was the mayor of St.-Pierre, who forbade anyone to leave town until important elections were completed.

Peleé erupted in a spectacular manner, emitting what is known in the vulcanology trade as a *nuée ardente*, a hot, heavy cloud of mixed gases and glowing particles. The cloud, heavier than the air, but turbulent enough to prevent settling of its material, spilled over the edge of the volcano and flowed down the mountainside. As such clouds do, it traveled at a speed of about a hundred miles per hour, right smack through St.-Pierre.

Within two minutes, the city of thirty thousand people was obliterated. The cloud scoured the area so thoroughly that even the layout of the streets could not be reliably determined afterward.

There was one survivor, a citizen who had spent the night in the drunk tank, a dungeon safely located in the depths of the prison. The outcome of the election remains uncertain.

A volcanic eruption itself is noted with interest and awe, although happy-go-lucky people dance on the rims of volcanoes that are already announcing their intention of erupting, and though people rebuild houses and schools and churches on the shoulders of volcanoes that have recently destroyed their houses, schools, and churches. The new volcanic soil tends to be extremely fertile once it has cooled, and the lush growth of green stuff soon puts everyone at ease. The time of the eruption is recorded as an interesting bit of history which carries no particular lesson for the future.

Still, the dates of eruptions are recorded fairly systematically and those dates are useful to us. (Do you remember the eruption of Mount Lamington in New Guinea in 1943? No? You were busy with other things at the time? The eruption killed twenty thousand people, according to some reports, and we find the information of interest statistically.)

Earthquakes are of indirect interest to us in this study, because they are closely associated with volcanoes and we can use the dates of reported earthquakes in our calculations of the Earth's response to tidal forces. To state this most simply without elaborate documentation: volcanoes and earthquakes seem to be different manifestations of the same process, the release of energy from the Earth's crust.

People notice earthquakes as much as they do volcanoes, and they comment on them in historical writings with great regularity. Every earthquake seems big to people at the epicenter and subjective reports of quakes are not always accurate indicators of their size.

It makes a lot of difference whether the quake occurs deep in the Earth or near the surface, far away or nearby. The amount of energy released is not the only criterion for judging earthquake intensity, but it's the best guide we have and energy release information is what we seek.

Though various scales are used by technical people to describe earthquake "size," we refer conventionally to the Richter scale, devised by Dr. C. F. Richter of Cal Tech in Pasadena. The scale runs from 0 through 8 and in-between steps are expressed in tenths, as "3.4 on the Richter scale," or "A whopping 8.3 on the Richter scale!"

Notice that a quake that is rated at 2 on the Richter scale is not twice as big in terms of ergs released as a quake rated a 1 on the

Richter scale. (Consult Table 2.1.) A Richter 1 quake releases about 7.9×10^{12} ergs, while a Richter 2 quake releases about 2.5×10^{14} ergs, *about thirty times more.*

It's a sort of strange scale, making these huge jumps from step to step, but at least it's tied to something that can be measured, the ratio between the amplitude of the pressure wave of a quake and the amplitude of the shear wave. Before that, the Mercalli scale of twelve was in use, based on subjective descriptions of how much crockery had been broken or how far rocks had been displaced.

When the 1971 San Fernando quake hit, the seismograph at Cal Tech was knocked off scale because it was practically on top of the action, so the newsmen all rushed over to Dr. Richter's house to ask him how big the quake was. After all, the man must know something about his own scale, what? Standing in his doorway, helpless but game, Dr. Richter explained that the quake was, well . . . uh . . . sort of . . . uh . . . very shaky! The reporters then announced excitedly that Dr. Richter had estimated that we'd had quite a severe earthquake. It wasn't exactly news by that time, but it made all the reporters feel better to have the word from the world's foremost authority.

Before Dr. Richter worked out his measuring system, seismometers had been around in reliable form for perhaps a hundred years and it was possible to convert their old records of raw data into the same scale we now use. Before that, however, we were stuck with purely subjective descriptions of quakes through history. We have no certain way to determine the power of ancient quakes. However, we do have means to make estimates.

Earthquakes tend to kill a lot of people and it is traditional to count the deaths and record them. Almost all societies feel it's important to account for dead people. We have reports of the numbers of deaths associated with a great many quakes. Consider Table 4.1. All of the data shown were extracted from standard references: *Encyclopaedia Britannica, Encyclopedia Americana,* the *World Almanac, Information Please Almanac, New York Times Encyclopedic Almanac,* and *Dunlop Illustrated Encyclopedia of Facts.*

Not all of the quakes are adequately described for our purposes (some dates and some estimates of size are vague), but this does give us a statistically useful number of quakes on which we have a fairly good handle.

Some of the reports of deaths are remarkable. In many cases, the

Table 4.1.

Great Earthquakes from Standard Reference Books

Date		Place	Deaths in Thousands
	526	Syria	250
	681	Japan	(3 sq. mi submerged)
	856	Greece	45
	869	Japan	(thousands)
	1038	Shansi, China	23
	1057	Chihli, China	25
	1170	Sicily	15
	1268	Asia Minor, Silicia	60
	1268	Po Hai (Gulf of Chihli)	100
27 Sept.	1290	Chihli, China	100
20 May	1293	Kamakura, Japan	30
	1361	Japan	(thousands)
	1456	Naples	30–40
26 Jan.	1531	Lisbon	30
24 Jan.	1556	Shensi, China	830
Nov.	1667	Caucasia, Shemaka	80
11 Jan.	1693	Catania, Italy	60
	1693	Naples	93
30 Dec.	1703	Tokyo	200
	1716	Algiers	20
7 Oct.	1737	Calcutta	300
7 June	1755	Northern Persia	40
1 Nov.	1755	Lisbon	60
30 Oct.	1759	Baalbek, Lebanon	30
4 Feb.	1783	Italy, Sicily	50
	1792	Japan	15
4 Feb.	1797	Peru, Ecuador	40
15 Dec.	1811	Missouri	?
23 Jan.	1812	Missouri	?
7 Feb.	1812	Missouri	?
5 Sept.	1822	Syria	22
28 Dec.	1828	Japan	30
	1841	Japan	12
	1847	Zenkoji, Japan	34
13 Aug.	1868	Peru, Ecuador	25
16 May	1875	Venezuela, Colombia	16
28 July	1883	Ischia	2
3 Aug.	1883	Tyrrhenian Sea	2
22 April	1884	England	?
31 Aug.	1886	South Carolina	.060
28 Oct.	1891	Japan	10
15 June	1896	Japan	27
	1898	Japan	22
18 Apr.	1906	San Francisco	.700
16 Aug.	1906	Chile	1.500
28 Dec.	1908	Italy, Sicily	100
13 Jan.	1915	Italy	30
11 Oct.	1918	Puerto Rico	.116

Table 4.1. *(cont'd)*

Great Earthquakes from Standard Reference Books

Date		Place	Deaths in Thousands
16 Dec.	1920	Kansu, China	180
1 Sept.	1923	Japan	200
23 July	1930	Italy	1.883
26 Dec.	1932	Kansu, China	70
3 Mar.	1933	Japan	2.500
10 Mar.	1933	California	.115
31 May	1935	India, Pakistan	50
24 Jan.	1939	Chile	40
27 Dec.	1939	Turkey	30
15 Jan.	1944	Argentina	5
1 Apr.	1946	Alaska	.173
28 June	1948	Japan	5
5 Aug.	1949	Ecuador	6
15 Aug.	1950	India	1.5
6 May	1951	El Salvador	1
21 July	1952	California	.014
21 Feb.	1953	Iran	1
18 Mar.	1953	Turkey	1.200
9 Sept.	1954	Algeria	1.600
10 June	1956	Afghanistan	2
2 July	1957	Iran	1.500
2 Dec.	1957	Outer Mongolia	1.200
13 Dec.	1957	Iran	1.392
29 Feb.	1960	Morocco	20
22 May	1960	Chile, Japan, Hawaii	2
1 Sept.	1962	Iran	10
26 July	1963	Yugoslavia	1.100
27 Mar.	1964	Alaska	.114
28 Mar.	1965	Chile	.400
19 Aug.	1966	Turkey	2.529
30 July	1967	Venezuela	.277
31 Aug.	1968	Iran	12
1 Oct.	1969	Peru	.136
28 Mar.	1970	Turkey	1
31 May	1970	Peru	30
10 Dec.	1970	Peru, Ecuador	.081
9 Feb.	1971	California	.064
12 May	1971	Turkey	.057
23 May	1971	Turkey	.800

deaths arose from secondary effects—well, almost *all* deaths are secondary effects when you come right down to it, since hardly anybody is outright shaken to death. Still, it seems not quite proper to count all of those 830,000 people in the Shensi, China, quake in

1556, since the greater part of the trouble arose from the accidental damming of a river, followed by the bursting of the dam.

It makes a deal of difference whether you live in a prestressed concrete building or an unreinforced adobe building when the shaking starts. The adobe structure tends to fall flat at once. This accounts for the great number of casualties in Managua, for example, or the large percentages of casualties in quakes in Iran.

That is to say, our record of quake magnitude indicated by deaths selects for populated areas and is not a comprehensive list of earthquakes around the world. For our purposes, however, this doesn't matter. We are not concerned much about quake location. We are far more concerned about quake *times*, and we are not aware of any bias with regard to times in this list. In addition, we assume with some confidence that the reports on casualties are distorted on about an equal, random basis. The reports taken as a whole are a reasonable basis for judging earthquake magnitudes, given that you have been forewarned that these are not 500-year-old seismograph records.

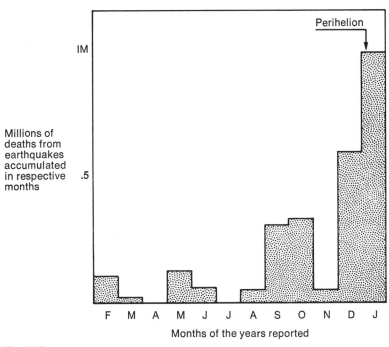

Fig. 4.10.

We made a major distinction among quakes, separating small ones from big ones, for reasons that will become clear. In the early quakes, for which we have no direct seismic records, a "small one" is a quake in which fewer than twenty thousand people are reported killed.

What we are working up to, of course, is the proposition that both earthquakes and volcanic eruptions are triggered by high tidal forces —not *caused* by tidal forces, but triggered by them. The straw that breaks the camel's back, and all that.

Consider first a monthly report of casualties. That is, the three million or so earthquake deaths on our list were divided up by the months in which they occurred.

Earthquake deaths do not occur randomly. They bunch up in December and January, close to perihelion.

Consider the eight greatest earthquakes of the twentieth century. Through 1968, these few quakes had released nearly 25 percent of the total energy released from the crust in this century. Notice that the renowned San Francisco quake of 1906 doesn't show on this list. It wasn't big by comparison with any of these quakes, two of which occurred in the same year—1906, a vintage year by all reports.

Table 4.2.			
Great Earthquakes from Standard Reference Books			
Date		*Place*	*Magnitude*
31 Jan.	1906	Colombia 1°N 82°W	8.6
17 Aug.	1906	Chile 33°S 72°W	8.4
3 Jan.	1911	Tien Shan Mts., Asia 44°N 78°E	8.4
16 Dec.	1920	Kanan, China 36°N 105°E	8.5
2 Mar.	1933	Japan 39°N 145°E	8.5
15 April	1950	China 29°N 97°E	8.6
22 May	1960	Chile 38°S 73.5°W	8.4
27 Mar.	1964	Alaska 61°N 148°W	8.5

Now, look at these same quakes plotted for both time of year and latitude.

Clearly, winter is heavily favored as a time for great earthquakes. The Northern Hemisphere is also favored. Why? Well, perihelion falls during winter in the Northern Hemisphere. That would account for the tendency to more quakes in winter there. Unfortunately, big quakes occur during winter in the Southern Hemisphere, too. So

Latitude vs. Time of Year

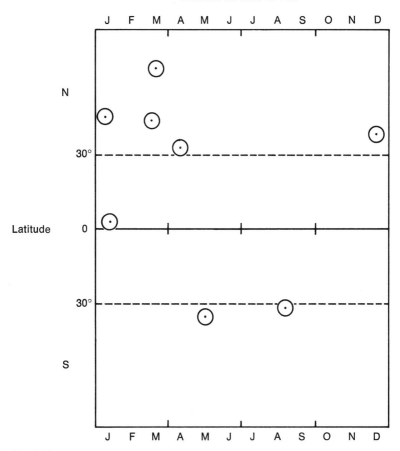

Fig. 4.11.

much for our rational explanation. This must simply be classed as an observation until we think of something.

A plot of quakes in which more than twenty thousand people were killed indicated that 13 occurred in winter

        5 occurred in spring
        3 occurred in summer
        6 occurred in fall
    ─────
    27

Winter wins again, hands down.

Perihelion appears to be a significant factor in the timing of large earthquakes.

To consider other factors, we have to look at a more complicated picture. Let us recall the piling up of tasks at your neighbor's house —or, more to the point, the summing of tidal forces on a periodic basis.

We have been concerned with perihelion; the next obvious item of interest is peri*gee*. Take a look at Figure 4.12, ignoring everything except the circular indicators around the slanted line at the bottom of the graph. Each of the circles is the time of one of those great earthquakes in this century. Importantly, the closer the circle is to the slanted line, the closer the earthquake was to perigee.

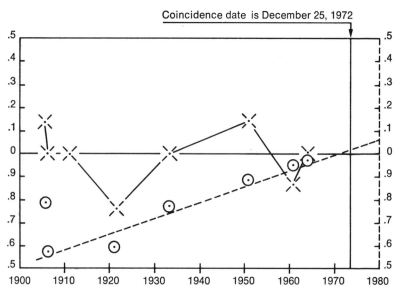

December 20.1, 1972, was highest tide
since December 19.3, 1964, at 30° N. latitude

Coincidence date is December 25, 1972

The vertical scales represent decimal fractions of constant-length time periods.

✕ Based on 4.425-year period of alignment of perigee with perihelion.

⊙ Based on 27.554551-day length of anomalistic month.

Fig. 4.12.

Clearly, seven of the eight quakes fell almost at the same time as perigee, and the eighth was off by only a small period of time. It does seem that perigee was significantly related to the timing of these great earthquakes.

What else?

We have talked about perihelion and perigee taken independently, but it is also true that they are related in their effect on tides on the Earth.

Every 4.425 years perigee and perihelion are aligned. That is, Sun, Earth, and Moon are close to being lined up when viewed from one angle *at the same time* when both perihelion and perigee are occurring.

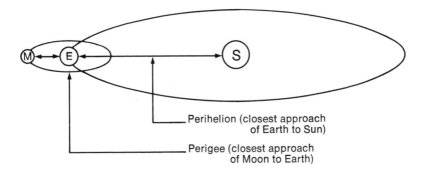

Perihelion (closest approach of Earth to Sun)

Perigee (closest approach of Moon to Earth)

or, 4.425 years later:

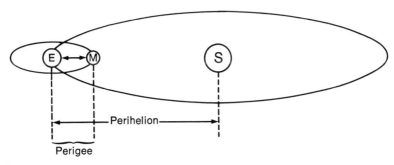

Perihelion

Perigee

Fig. 4.13.

This doesn't mean that an eclipse occurs at the same time (although that happens, too, at longer intervals), since the alignment may be far off when the system is viewed from a different angle.

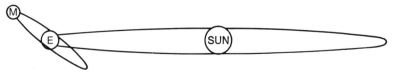

Fig. 4.14.

The 4.425 years is one-half of the full 8.85-year "perigee precession cycle" since the position of perigee has moved just around to the other side of the Earth in 4.425 years, not clear around 360 degrees and back to where it started.

The question, then, is whether or not the great earthquakes we have been using as examples have occurred at particular times related to a 4.425-year cycle. Look back at Figure 4.12.

The "x's" indicate the occurrence of those quakes with respect to the perigee precession semicycles. That is, the closer the x is to the horizontal line in the middle of the graph, the closer it was to *alignment of perigee and perihelion.*

Three of the x's fall right on the line; one is just slightly off the line; the other four are significantly close to the line. Are perigee/perihelion alignments significantly related to the timing of great earthquakes? It looks like it to us.

We have been ignoring a lot of the additional information on this graph—not because it doesn't matter, but because it is so difficult to explain. If you are really determined to learn about a couple of the more complicated calculations, you will find them discussed in a brief appendix at the end of this chapter.

Consider another measure of the relationship of the 4.425-year period to earthquakes. We have already listed a number of quakes whose magnitude can be estimated only by the reports of deaths they caused (Table 4.1). We do have the dates of most of those quakes, too.

Suppose we pick some arbitrary reference date and and count back in increments of 4.425 years to see if those quakes occurred at any particular time with respect to such intervals. (This method is discussed in the appendix.)

This will tell us something more about those quakes, as it did about the eight big ones in this century. Note that we distinguish between "big" quakes killing more than twenty thousand people and "little" quakes killing fewer than twenty thousand people. Our theory is that tidal stresses trigger big quakes, but have a much less significant relationship with small quakes.

Figure 4.15 indicates a very significant relationship between perigee/perihelion alignments and big earthquakes, a less clear relationship with small quakes.

The lower graph lumps large and small quakes together. The distribution of earthquakes with respect to the 4.425-year perigee precession semicycle is hardly what you'd call random.

We have not yet discussed the "eclipse node semicycle," which gives us the best possible alignment of Earth, Sun, and Moon every 9.3 years.

Figures 4.16 and 4.17 show a time sequence analysis based on the 9.3-year cycle measured from the reference point of 1 January 1972 (a point very close to perihelion).

Figure 4.16 includes "small" quakes killing more than ten thousand people; Figure 4.17 shows large quakes killing over twenty thousand people.

Interestingly, the smaller quakes (and we don't regard ten thousand deaths as all that small) seem to indicate a more nearly periodic performance in these plots, but periodicity is obvious in both cases.

How does this period add up with the 4.425-year cycle? You can plot a number of periods that will fit both of these. When you find a period into which both 9.3 and 4.425 fit an even number of times, you've found a positive, reinforcing period—one of those times when the neighborly tasks pile up. On the other hand, of course, you may identify quiet periods during which the forces are consistently low, the factors being out of phase. The good neighbor did have four hours of unbroken rest every once in a while.

We are persuaded that tidal forces trigger major earthquakes. If a really high tide is coming (e.g., February 1971) you may reasonably expect quakes to occur here and there around the world at the same time—though not necessarily right on the hour or even on the exact day of the highest tidal force.

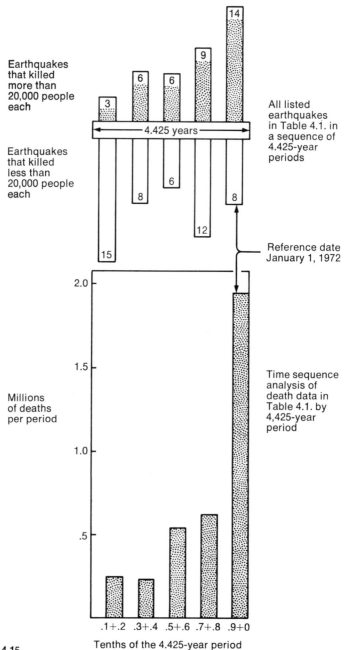

Earthquakes that killed more than 20,000 people each

Earthquakes that killed less than 20,000 people each

All listed earthquakes in Table 4.1. in a sequence of 4.425-year periods

4.425 years

3

6

6

9

14

8

6

12

15

8

Reference date January 1, 1972

Millions of deaths per period

2.0

1.5

1.0

.5

Time sequence analysis of death data in Table 4.1. by 4,425-year period

.1+.2 .3+.4 .5+.6 .7+.8 .9+0
Tenths of the 4.425-year period

Fig. 4.15.

70

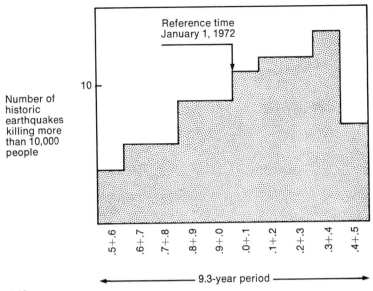

Number of historic earthquakes killing more than 10,000 people

Reference time January 1, 1972

10

.5+.6 .6+.7 .7+.8 .8+.9 .9+.0 .0+.1 .1+.2 .2+.3 .3+.4 .4+.5

←————— 9.3-year period —————→

Fig. 4.16.

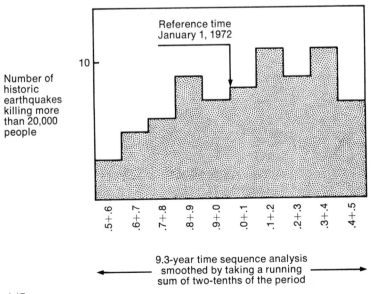

Number of historic earthquakes killing more than 20,000 people

Reference time January 1, 1972

10

.5+.6 .6+.7 .7+.8 .8+.9 .9+.0 .0+.1 .1+.2 .2+.3 .3+.4 .4+.5

←————— 9.3-year time sequence analysis smoothed by taking a running sum of two-tenths of the period —————→

Fig. 4.17.

71

As the alignment approaches its peak, the tidal force is rising sharply. One never knows how much force is enough. The triggering is likely to occur at any point, not necessarily waiting for the peak. The theory on which we pointed to 6 A.M. or 6 P.M. for the San Fernando quake is that *lateral* tidal force would be greatest right along the terminator, the edge between darkness and light at dawn and dusk.

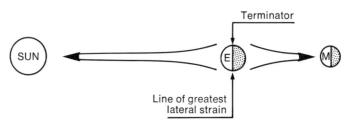

Fig. 4.18.

Further, for reasons that are not clear to us yet, there seems often to be a delay in the timing of the quake so that the excitement breaks out even three or four days after the peak force has passed. Perhaps there is some propagation delay, some gathering of forces, or a recoil. Maybe the reasons are perfectly obvious, but not to us at this time.

While we are still discussing earthquakes and tides, let's mention moonquakes.

One of the delightful things about the Apollo missions was that the astronauts left seismic stations on the surface of the Moon which report on quakes. The Moon is a great place to make seismic measurements, since the signals come through so nice and clear. There are comparatively few trucks driving past the station to rattle the instruments and the general level of background noise is very low. This makes it possible to sort out very tiny discrete events in the records.

Latham and his associates[1] reported on the lunar seismic experiments in 1971, pointing out that accumulated strain in the Moon is apparently triggered by tidal forces. Indeed, the thing that sticks out of the data is the fact that every time the Moon passes through perigee, a whole swarm of little quakes occurs.

1. G. Latham et al., "Moonquakes," *Science,* 174 (1971): 687, 692.

We haven't had time to accumulate data on the 4.425-year period with respect to moonquakes, but some of us look forward to that information with great interest. Meanwhile, we are tickled to have something directly useful come to us so quickly from the Apollo program.

And, still on earthquakes . . .

Tidal forces can only *trigger* quakes, not provide the energy that they expend. Knowing when the tides are due is just part of the game in earthquake prediction. For example, we were able to guess in advance that the week before Christmas, 1972, was a time of high earthquake probability, owing to good Sun, Earth, Moon alignments near perihelion. Indeed, we fingered the period from December 20 to December 25 as the time of maximum danger—and the Managua quake fell smack in the middle of that period, on December 23. (Actually, the Managua quake was not especially powerful, but a city built largely of adobe tends to fall heavily on its inhabitants without much provocation.) However, we had no way of knowing *where* a quake in that period might be triggered, apart from the general expectation that quakes will occur in "earthquake zones."

Similarly, we had no reason to suppose that a quake would occur in the San Fernando Valley instead of in San Francisco or Athens or Santiago.

Times are changing, however. In 1973 some chaps at Lamont-Doherty Geological Observatory published their elegant work on what they call the "dilitancy theory," which was stimulated largely by observations reported by Soviet scientists. (There is reason to hope that this Soviet research might contribute ultimately to saving American lives when we have learned to predict earthquakes accurately.) In a nutshell, the apparent fact is that the relative transmission speed of seismic P waves and S waves changes noticeably in an area that is building up to a quake. The pattern of change may be observed readily over a period of days, weeks, even months. The reported observations and correlations with the occurrence of real quakes are very impressive.

It is interesting to look at the plots that have been prepared of the seismic transmission data records before the San Fernando Valley quake of 1971. If we had known of the existence and the significance of these signals at the time, we'd have had not only a pointer to the time of triggering, but a real guide to the probable quake location.

We might have pretended to be sick and have put off our consulting date until a more propitious time, say when perihelion and perigee were farther out of phase.

Volcanoes are our primary concern here, of course, and we find that volcanic activity does follow the same periodic schedules that the earthquakes follow.

An English climatologist, H. H. Lamb, has for some years published interesting papers examining the relationship of volcanic dust in the atmosphere to surface temperature on the Earth. He published a very useful, if incomplete, list of volcanoes along with a numerical estimate of the amount of dust which each contributed to the atmosphere. This is referred to as the "dust veil index."

(To say that the list is incomplete is not to complain, mind you. It is a massive labor to gather such material as this, and Lamb's labors are not lost. His list and estimates are immensely useful and we view with mixed dread and enthusiasm the opportunity we see to add several hundred volcanic eruptions to the list. Hooray for Lamb and we wish he'd collect all the others for us.)

Lamb's list runs from 1500 through 1963, with ninety-eight entries. The number is statistically significant. With a time span approaching five hundred years, we have a nice, long slice of history in which to do time sequence analysis. Such analysis reveals the relationship, if any, of various events to particular periods of time.

Figure 4.19 shows the results of a 4.425-year time sequence analysis based on Lamb's dust veil estimates. This plot was made before it occurred to us that 4.425 years was the key number, rather than the 8.85 full perigee precession cycle. Indeed, it was a look at this chart with its outstanding 4.425-year pattern that forced the realization that the semicycle was more significant than the full cycle. Figure 4.20 is based on the full cycle.

It is difficult to reject the notion that volcanic activity is powerfully influenced by cyclic tidal forces.

In 1972 Johnston and Mauk[2] reported on their observations of Stromboli (to pick an interesting volcano at random) and their comparison of its very frequent eruptions with the tidal forces which can be determined from Longman's formulas.

2. M. J. Johnston and F. J. Mauk, "Earth Tides and the Triggering of Eruptions from Mount Stromboli, Italy," *Nature*, 239 (1972): 266.

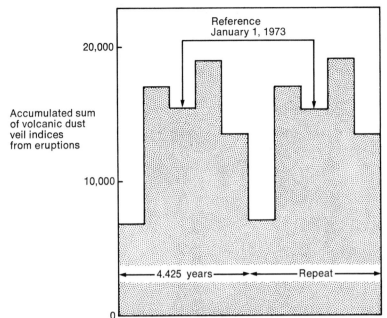

Fig. 4.19.

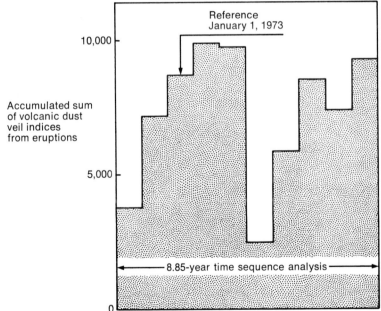

Fig. 4.20.

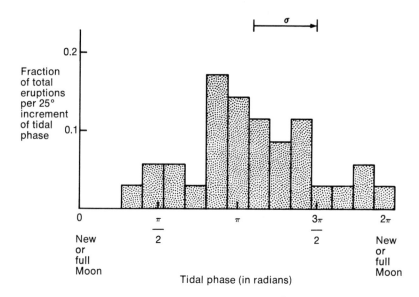

Fig. 4.21.

After Johnston and Mauk 1972

Sure enough, Stromboli appears to pulse with the fortnightly tides in convincing style.

For some frosting on this cake, consider a report on the work of John Rinehart of the University of Colorado in Boulder, who studied the performance of three geysers and searched for regular periods in their spouting schedules. One of the major components of the schedules appeared to be the fortnightly tide. Further, the geysers seem to respond to the same forces that tense things up preceding earthquakes.

Rinehart's plot of average interval between eruptions in his northern hemisphere geysers indicates that over a period of about seven and a half years, the interval between eruptions is consistently longest early in the year, dropping to minimum at midyear, and rising again toward maximum at the end of the year with the approach of perihelion.

We don't necessarily suggest cause and effect here. This is just a report on the man's observations.[3]

3. To those who have little confidence in statistical calculations of this sort, we can offer little reassurance. However, we point out that conventional statistical techniques were used in our work to determine the probability that

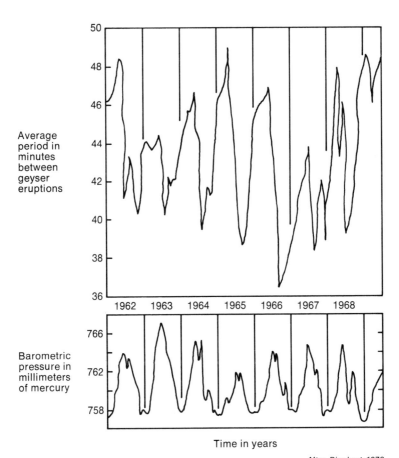

Average period in minutes between geyser eruptions

Barometric pressure in millimeters of mercury

Time in years

After Rinehart 1972

Fig. 4.22.

any particular calculated value is significant. That is, what is the probability that a particular answer is the product of chance, a fluke, coincidence?

The standard chi square (X^2) is applied for each list of figures against a null hypothesis except in cases where the number is low (i.e., less than 25 or 30). The chi square test is not a valid test in the case of data being progressive —as in a sawtooth frequency change. In this case, the *sequence* of increasing numbers is also significant, hence a rank correlation technique is required.

For both low numbers of events and rank correlation, instead of running all of the permutations, the *worst* case was taken for the null hypothesis and tested by chi square.

The values stated are taken linguistically as follows:

77

A dry summary of all this:

1. Large earthquakes are triggered by the additional tidal forces caused by the *perihelion* configuration.

   P is less than .05, a significant relationship.

2. Great earthquakes tend to occur in *winter*.

   P is less than .05, a significant relationship.

3. Moonquakes occur at *perigee*.

   P is much less than .001, an extremely significant relationship.

4. Great earthquakes occur at *perigee*.

   P is much less than .001, an extremely significant relationship.

5. There is earthquake triggering in synchrony with a tidal force beat due to the alignment of *perigee and full or new moon* every $413+$ days, and there is an extremely significant probability of occurrence in the northern and southern hemispheres 180° out of phase.

   P is less than .001

   $29.530589 \times 14 = 413.428246$ days

   $27.554551 \times 15 = 413.318265$ days

   The tidal force beat is reflected in the "sawtooth" frequency of earthquakes since 1900.

6. There is a tidal force beat due to alignment of the *perigee* with *perihelion*, which happens at 8.85/2-year intervals (i.e., 4.425 years), which affects earthquakes and dust veil triggering.

   P is less than .001, an extremely significant relationship.

7. Great earthquakes occur in synchrony with a tidal beat every 9.3 years, i.e., coincidence of *perihelion* and *eclipse* nodal alignment.

8. A significant outburst of great volcanic eruptions occurs in synchrony with the approximate coincidence of *perigee, perihelion*, and an eclipse node.

   P is less than .05, a significant relationship.

P = .05 is significant.
P = .01 is very significant.
P = .001 is extremely significant.
That is to say, if the values tested could have been expected to deviate from random as much as they did only once in a thousand times, then the values were extremely significantly different from random (the null hypothesis).
This methodology is all standard statistics. If you are convinced that statistics is just sleight-of-hand, then this won't make you feel any better.

Our suspicion, based on something more than native intuition, is that tidal forces *do* govern the activity of volcanoes.

APPENDIX

Some of the calculations upon which our conclusions in this chapter were based deserve brief explanation. For example, Figure 4.12, showing the relationship of the anomalistic month and the perigee precession semicycle to the timing of the great earthquakes of this century, was based on examination of Table 4.3.

Table 4.3.

The Eight Great Earthquakes of the Twentieth Century

(As listed by the Encyclopedia Americana)

Date	Decimal Equivalent of Date	Years Ref. 12/25/72 1972.9856473	Years 4.425	Years Anomalistic Months
1/31/1906	1906.0848752	66.9007721	15.119	886.786
8/17/1906	1906.6297191	66.3559282	14.996	879.564
1/3/1911	1911.0082137	61.9774336	14.006	821.526
12/16/1920	1920.9610062	52.0246411	11.757	689.599
3/2/1933	1933.1697504	39.8158969	8.998	527.769
4/15/1950	1950.2874805	22.6981668	5.129	300.870
5/22/1960	1960.3915210	12.5941263	2.846	166.938
3/27/1964	1964.2381981	8.7474492	1.977	115.950

Note: Ignore the number of cycles and plot the decimal fractions, i.e., the phase.

Table 4.3 uses Christmas Day, 1972, as a reference point.

The date of the quake is first given in standard notation.

The date is next given in a decimal equivalent.

The next entry is the number of years before the reference date at which the particular quake occurred (again in decimal equivalent).

The next entry is the number of times that 4.425 years goes into that differential number of years from the reference date.

The last column indicates the number of anomalistic months in that same differential number of years from the reference date.

So what?

What we're trying to find is alignment of perigee and perihelion. It would be significant if perigee and perihelion were found to align or to be very *close* to alignment on the date and at the hour of each of the great quakes.

The trick is to look at the numbers *after the decimal point* in the last columns. If these numbers show that the entry is quite close to being a whole number, then the alignment is good. If the numbers after the decimal point indicate that the entry is not close to being a whole number, then the alignment is not good—the phenomena are "out of phase."

When both the 4.425-year value and the anomalistic month value are close to being whole numbers, then we know that the tidal forces associated with both perigee and perihelion were both nearly maximized at the time of the quake.

In fact, the numbers are all very close to being whole numbers.

Figure 4.12 indicates this effect more clearly. Four of the 4.425-year calculations fall almost on the whole number line at "0." The other four entries are plus or minus the whole number by no more than 0.243 and in most cases much closer to the whole number line than that.

The anomalistic month calculations are much the same. For reasons that do not matter for this purpose, the whole number line for the anomalistic months slants up across the graph.

The critical point here is that perigee and perihelion were in very good alignment, providing comparatively great triggering tidal forces on the dates of those great quakes.

The 4.425-year perigee/perihelion alignment period seems to be one of the very basic units of which our tidal cycles are composed. This is like your requirement for taking a pill every four hours while tending the neighbor's pets; the period shows up clearly in the complex series of events of which it is a part. We are confident, if we divide an apparent cyclic period by 4.425 and find that the answer is an even number or very close to it, that the perigee precession semicycle is a significant factor in the apparent cyclic period.

Figure 4.15, showing earthquake distribution (large and small) with respect to the 4.425-year cycle, was generated this way:

First, we established January 1, 1972, as a reference date, very close to perihelion. Then, for each of twenty-eight large quakes, we counted back from our reference date in 4.425-year increments. The actual date of the quake, then, fell into some fraction of the last 4.425-year period. That is, the quakes didn't all fall exactly at the end of an increment, but usually somewhere inside that last increment. We divided that increment into five sections.

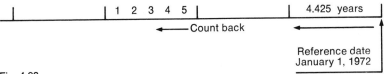

Fig. 4.23.

In this example, the actual date of a quake was in the very middle of the fourth 4.425-year period back from the reference date. Now, we don't care how long ago the quake was, but we do care about its place inside that last period. Since it was in the third section, we tallied that quake in the third column of our graph. Other quakes, of course, fell into different sections of their periods.

Indeed, as you can see from Figure 4.15, the distribution for large quakes was nonrandom. The closer you get to the end of the period, the closer you are to perigee/perihelion alignment. For large quakes, obviously, there is some important relationship with the 4.425-year period.

This treatment of the data is called a time sequence analysis. It is the sort of thing you do if you are trying to figure out whether there is any periodic factor in a big mess of random-looking numbers scattered through time. We did it and there was.

5

We have gone to great pains to point out that volcanoes seem to be triggered by tidal forces. Why do we care? If you don't live on a volcano, why should it interest you personally?

We have mentioned the proposition that bolide strikes throw large quantities of dust into the atmosphere and the dust forms a screen that prevents warming sunlight from reaching the Earth.

Volcanoes do the same thing, though the mechanism is not quite so simple. This is the effect that H. H. Lamb has been working with and which interested Franklin.

The indications of volcanic dust veil are widespread in the literature, running to hundreds of pages, but we'll select here a few pointed reports that seem to be representative.

In a 1913 paper Abbot and Fowle noted sharp reductions in solar radiation which they attributed to "volcanic effects"—in 1885, 1891, and 1903.[1]

They noted the effects of the 1912 eruption of Mount Katmai in Alaska on solar radiation measurements at Mount Wilson in California and in Bassour, Algeria. A table "Illustrating Opacity Due to Volcanic Matter" indicates the decrease in solar radiation from early June to August.

The text comments: "The maximum decrease in solar radiation

1. C. G. Abbot and F. E. Fowle, "Volcanoes and Climate," Smithsonian Miscellaneous Collections, vol. 60, no. 29, 1913.

attributable to the haze seems to have reached nearly or quite 20 percent at each station."

Roosen, Angione, and Klemcke reported in 1972 on their studies of atmospheric transmission of radiation. They commented: "Except for sporadic perturbation due to volcanic activity, there has been no detectable change in the global atmospheric transmission measured from remote sites in the last half century."[2] And they documented their work handsomely.

In 1972, Karen Heidel reported on "Turbidity Trends at Tucson, Arizona."[3] To oversimplify the matter, turbidity measurements indicate the amount of dust in the atmosphere, the key measurements being made by a pyroheliometer, which reads solar intensity.

The plot of turbidity at Tucson for the years from 1957 through 1971 show modest variation until 1963, when Mount Agung in Bali, a big duster, erupted.

From that point, the turbidity in Tucson increased dramatically, crested in the middle of 1964, and drifted back down. Incidentally, there are big copper smelters near Tucson, about which the conservationists are wont to worry. Heidel notes two smelter strikes during the observed period, during which operations were shut down for several months at a time. She concludes that the smelters "do have a small significant effect on Tucson's turbidity. . . ." It's interesting that a volcano thousands of miles away should have a *huge* effect.

Gow and Williamson reported a piece of work in 1972[4] which capped the previous literature and significantly influenced our own examination of this field.

They were involved in a task to which we'll refer more as we go along, the examination of ice cores bored out of the thick ice cover of Antarctica. The ice occurs in layers which can be dated back through many thousands of years. Special analysis techniques permit

2. R. G. Roosen, R. J. Angione, and C. H. Klemcke, "Worldwide Variations in Atmospheric Transmission, I. Baseline Results from Smithsonian Observatory," unpublished paper.

3. Karen Heidel, "Turbidity Trends at Tucson, Arizona," *Science*, vol. 177, 1972.

4. "Eruptions and Climate—Proof from an Icy Pudding," *New Scientist*, Feb. 17, 1972: 367.

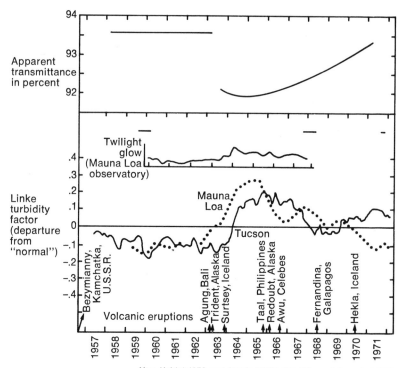

Fig. 5.1.

After Heidel 1972 and Cronin 1971; also Ellis and Pueschel 1971

determination of the temperature at which each layer was deposited, so the cores may be used as recordings of the climate through the millennia. In addition, the researchers discovered layers of volcanic ash at certain levels. The caption next to one of the illustrations makes our point here:

> A decline in Antarctic temperature of 2° to 3° beginning some thirty thousand years ago, coincided with the onset of prolonged volcanic ash deposition. The lowest temperature, sixteen to seventeen thousand years ago, matches the peak fallout of ash; and a warming follows the cessation of volcanic activity.

Commenting on the work of Gow and Williamson in an article titled "Eruptions and Climate: Proof from an Icy Pudding," the *New Scientist* says:

Another calculation implies that increasing the dust concentration of the Earth's atmosphere by four times could decrease temperatures by 3–5°C and even bring on an ice age if the condition persisted for a number of years.

. . . It seems that widespread volcanism can have more than the short-term climatic effects previously observed.[5]

It is astonishing to realize that just a few degrees of temperature make the difference between an ice age and tropical forests. This is not to say that it is a small thing to change the average surface temperature of the Earth by 5° or 6°F. Anybody who buys fuel oil to heat a house in the winter knows what enormous expenditures are required to keep temperatures comfortable for human beings even in a very small, confined area.

Think in these same terms of warming and cooling the Earth as a whole. ("Children, shut that door after you! We can't afford to heat the whole outdoors!") There's a lot of energy involved, but considering daily fluctuations of forty degrees in what we choose to call the "temperate" zones, a few-degree change in average temperature seems not very impressive.

What brings all this about? How much dust does a volcano put up?

The simplest obvious mechanism is the trick some volcanoes have of throwing fine particulate dust into the stratosphere. Agung, in 1963, seems to have fired about six-tenths of a cubic mile of material into the upper atmosphere this way, while Krakatoa in its famous 1883 eruption put up about 1.25 cubic miles of material. For all its renown, Krakatoa was not the biggest volcano anybody can remember. Lamb lists at least eighteen eruptions as large as or larger than Krakatoa since 1500. Coseguinta, in 1835, put up about *four times* as much material in one impressive blast.

The effect on world temperature is a function not simply of occasional large eruptions, but of the average dust veil produced by a large number of active volcanoes. The stuff doesn't drop back to earth in a few weeks, either, but stays aloft for some years, gradually dissipating.

The particles we're talking about are little guys, ranging from perhaps half a micron to three microns in diameter. You could line

5. Ibid.

up about fifty thousand half-micron particles in one inch along the edge of a ruler. A cubic inch could contain about 1.25×10^{14} of these particles. A cubic mile is a whole lot of cubic inches and that's a whole lot of particles to distribute around the world.

The particles not only reflect the incoming solar radiation, but gradually drift down through the atmosphere. Each particle has some probability of seeding a drop of rain. Thus, we not only have a temperature drop, but an increase in precipitation owing to the dust.

While this direct scattering of dust is obvious, another effect is not so obvious and it may be more important. Volcanoes pump out large amounts of sulphur dioxide—SO_2. As a matter of fact, the big dusters are comparatively rare, compared with the number of volcanoes that vent SO_2 into the atmosphere. It isn't just a matter of jetting the stuff into the stratosphere (and we keep making this point about the stratosphere, because the lower level material is apparently not very significant, dropping out comparatively soon.) Lava particles thrown to high altitudes tend to emit sulphur dioxide and the stuff is boiling off all over the place around volcanoes. It's a transparent gas.

Within recent years it has been noted that ultraviolet light reduces SO_2 to its components, sulphur and oxygen. Ultraviolet radiation is very heavy above the ozone layer, which filters it out before much of it can reach the surface of the Earth.

Another type of reaction which SO_2 is postulated to undergo in the stratosphere leads to ammonium sulfate, which also acts as dust.

Observers wondered for many years about the origin of "noctilucent clouds," which are often visible after dark at extremely high altitudes—as much as fifty miles, well above the ordinary cloud-forming zones of the atmosphere. The clouds seem to glow because they are lighted by the Sun, which has already dropped below the horizon of the viewer, but which still lights those very high clouds "from underneath."

Some workers have now concluded that these clouds are formed of the sulphur produced by the reduction of the SO_2 and/or the ammonium sulfate. An odd thing about the clouds is that they are durable. They last and last, apparently, because over a long time they are being formed about as fast as they dissipate. As the particles drift down from the clouds, still more are being formed above.

And the particles *do* drift down. Again, they are about a micron in diameter, but they have a consistent crystalline shape that is far more regular than the particulate rock of which we spoke earlier. Indeed, the crystal shape is much like the crystalline structure of ice, and this makes the particles ideal for seeding the precipitation of water vapor.

The quantity of SO_2 emitted by volcanoes is the subject of some discussion, especially since conservationists are distressed about man's contribution to change. Some efforts are now being made to figure out how great that contribution is, compared with other things.

Stoiber and Jepsen[6] estimate very conservatively that 10 million tons of SO_2 are pumped into the atmosphere by volcanoes annually, making the contribution not significantly larger than what some people have calculated as man's contribution to the atmosphere. They point out, however, that they had made their calculations on the basis of only a hundred current vents, that the measurements did not account for all the output from very large vents, that the calculations did not include the output of sporadic great eruptions, and did not include volcanoes under the sea (which has an area six times that of land). The sum of all these other inputs seems to us more likely to exceed 100 million tons per year of SO_2 than the 10 million.

Exposed hot lava evaporates and condenses in the air as an aerosol.

Carbon dioxide, too, is emitted in volcanic eruptions. Whereas the carbon dioxide has a warming (greenhouse) effect, the aerosols overcompensate and produce cooling. The exact effect of aerosols on temperature is highly controversial, but the gross effects of great haziness due to eruptions are clear: the Earth grows cooler after large eruptions. Enough eruptions will produce an ice age.

While we can do all sorts of interesting things with graphs to smooth them and to show averages (indeed, we *must* do this so that we can see the forest in spite of the trees), the volcanic events we plot are not themselves nice and smooth. Volcanic action doesn't just purr along; it goes Bang! Bang! The effects of volcanic eruptions don't show up in detail as smooth changes in averages, but as fairly sharp discontinuities. The Agung eruption sticks out like a sore thumb in Figure 5.1. We'd expect sharp changes to occur in other records, too, and they do.

Nature has provided several different means to record the events of the past, and people are growing increasingly skillful in reading those

6. R. B. Stoiber and A. Jepsen, "Sulfur Dioxide Contributions to the Atmosphere by Volcanoes," *Science*, vol. 182, 1973.

records. We have been particularly concerned with tree rings, ice cores, varves (cores from the bottoms of oceans and lakes, especially those that have no outlets), and geological cores.

Tree rings first.

Dendrochronology, the study of the growth rings in trees to determine and date past events, was not brand new when an American scientist, Andrew E. Douglass, turned his attention to the matter in the 1920s. However, Douglass applied himself with such vigor that he became the grand old man of the field in his lifetime, laying the groundwork for studies that may be carried on indefinitely.

All of us heard in school about how tree rings grow thicker in good years and thinner in bad, so that one can obtain a running weather report back through as many years as he has rings. Most of us retain a simpleminded picture of clearly defined thick and thin rings in a slice of a tree trunk. When we get down to the real thing, of course, the rings don't seem to be much different from one to the next and we have to take careful measurements with instruments. The trees often grow in a sort of lopsided fashion, so that a ring that is thick on one side is thinner on the other. Occasionally, two rings will develop for one year, while in some years, no ring develops. To spoil the rest of the fun, we seldom get nice slices of tree trunks like giant pickle chips. More often, the researchers have a core that has been taken out of the wood, a plug which, with any luck at all, contains samples of the rings from the outside of the tree to the very center.

The fact is that no slice of a tree means very much by itself. Only when the information gained from a particular slice is correlated with other information and averaged with other sections from its area does it give us any sort of message.

What sort of message are we looking for?

Douglass was trying, among other things, to establish the dates at which a large number of archeological structures were built. Specifically, the Indian pueblos of the Southwest were of interest. This area is littered with abandoned buildings of adobe, stone, and timber, whose origins have been obscure into quite recent times.

Well begun on his tree-ring studies, Douglass took cores of the timbers from these abandoned pueblos and examined them. To make sense of what he found, he also cored timbers in occupied pueblos whose dates of origin were known. This may sound sterile and scientific, but it involved delicate negotiations between Douglass and the

Indian leaders, who could see no good reason to let some crazy pale-eyes into their very private buildings to pry into their secrets. There was no small element of personal danger as Douglass took his little auger into hostile territory. Occasionally he was required to plug the holes with turquoise as a gesture of goodwill. Douglass became not only a gold mine of scientific information, but a treasure-trove of great stories.

Douglass was trying to piece together a consistent tree-ring sequence that would let him date not only the pueblos, but any other large hunk of wood that could be found in the area.

The fact is that few trees live for thousands of years. The California redwoods and the bristlecone pines (also California-based) do happen to last for a long time and it is good fun to examine a huge redwood slab on which the dates of particular rings have been identified. "This tree was a three-inch sapling when Caesar was assassinated by his friends. When Christ was born, this was a sturdy young tree six inches thick. When Charlemagne was crowned . . ." right up to the time a logging crew appeared in the forest.

Suppose, though, that you're dealing with trees that live for only a couple of hundred years. Suppose further that you are not on the logging crew that cuts the thing down, establishing clearly the date of the outermost ring, but instead you find this piece of wood built into the roof of an adobe building in a remote canyon where people haven't lived for hundreds of years. It's all very well to admire the rings, but how do you fasten them to a familiar calendar?

With enormous difficulty, that's how.

Imagine that you take your hatchet right now and you go out and chop down a tree. Look at the rings. Plot their pattern of thickness.

Hunt around, now, looking in old sheds or back yards, trying to find things made out of trees that were cut in the same area at earlier times. Say that you are either persuasive or a skilled burglar and you obtain from your neighbors a number of likely pieces of wood. You slice them and examine their ring patterns. Your face lights up when you work through one of them because you find what you are looking for: a series of rings whose pattern matches a section of the tree you chopped down yourself.

You can be reasonably sure that the patterns match because you have such a large number of data points to compare . . . all those rings. Notice that the rings that developed in your tree when it was

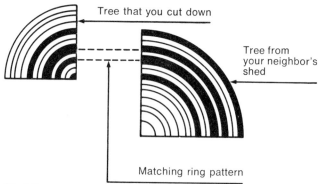

Fig. 5.2.

but a sapling match the rings in the other tree when it was well grown. The earlier rings in the older tree thus carry your record of rings back in time quite a number of years.

Now, talking with the oldest inhabitant in your town, you weasel out of him information about where you might find even older pieces of wood that were cut in the same area. It turns out that the lintel over the door of the city hall was cut from a tree that grew on the site. The old-timer remembers when the hall was dedicated and what a big fuss was made about the log from the good old tree that used to stand on the village green.

After a six-month struggle with the bureaucracy, you manage to get permission to take a core from that piece of wood over the front door of the city hall.

The tree-ring plot of this core extends your records back an additional two hundred years, perhaps.

Then out you go for another log somewhere—maybe furniture in the local history museum, a hunk of the old railroad station, a very ancient fence post. . . .

Obviously, this gets to be a formidable task after you have moved back just a few generations. You may lose the thread. Far more frustrating, you may turn up some slices that are older than anything you have previously found. Indeed, you may at a single archeological site find a lot of different pieces of wood which you can connect together in an unbroken sequence lasting for three hundred years—but you can't hook either end of that sequence to anything else. Douglass developed exactly such floating sequences. He spent years tracking down pieces of wood that filled in the gaps. It was a remarkable labor that he did.

After these many years since the 1920s, when Douglass was doing his basic work, his successors have managed to establish unbroken sequences of ring patterns which run back several thousand years in certain areas. This has made it possible to examine wooden artifacts found at archeological sites and determine the earliest possible dates at which they might have been made.

These tree-ring plots have proved enormously useful and they are being extended and expanded constantly to provide us with better worldwide dating techniques.

Again, this was but one of Douglass's interests. He was very much concerned with climatic history and with the possibility that tree-ring patterns would reveal cyclic effects.

Let us recall your chores with neighbor's pets for a moment. We were interested chiefly in the pattern of activity which developed from a given set of scheduled tasks. We started with the periods of

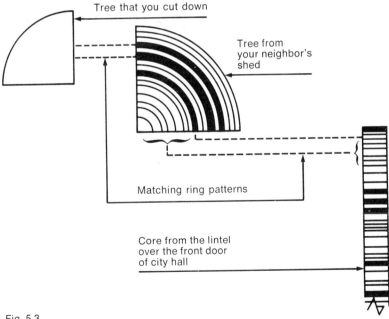

Tree that you cut down

Tree from your neighbor's shed

Matching ring patterns

Core from the lintel over the front door of city hall

Fig. 5.3.

the tasks and developed them into a composite pattern. We could have started with the composite pattern and determined from it the periods from which it was constructed. This was the situation of Douglass. He had the composite patterns of tree rings over hundreds of years and he searched those patterns to find out if they were built up of regular periodic components.

Douglass even built an optical analyzer that helped him to process his data. He did find apparent cyclic patterns, a good many of them, and he published his findings. (We are lucky enough to be acquainted with Major General Roy Lassetter, retired, who was a student of Douglass in . . . well, some years back. General Lassetter has lent us a lot of Douglass's publications and has given us a feel for the work which we could not otherwise have gained.)

In the years since Douglass died, people with computers have re-analyzed his work and decided pretty much that the old man missed the boat. *Sic transit gloria mundi.*

Luckily, though, Douglass published not only his conclusions, but the data upon which they were based. This allows us outsiders to go back and try our own calculations on the raw material. At least a few of us do find regular, periodic cycles in the raw material. Obviously, we're using a somewhat different approach from that of the no-cycle school. We suspect that the difference in results is produced by the difference in the size of the units that are used. If, for example, you take averages of tree-ring growth by decades, you are not likely to detect fourteen-month cycles. The small stuff disappears into the longer term averages.

Similarly, if you consider periods of a hundred years at a time, broken into units however small, you are unlikely to detect cycles longer than a hundred years.

The age of the tree examined has a great deal to do with the significance of its rings, too. Tree rings are living things when they form and they represent reality as the tree detects it at the time. A young tree is like a young person—every event seems all-important. If a deer were to hook a small tree and loosen its roots, the rings might later report that year as *terrible*. A small tree located in the bottom of a stream bed might report unceasing wet weather during a period of drought. The same tree, grown to appreciable size, might spread its roots both deep into the moist soil and off into the dry area away from the bottom of the stream bed. Its reports on the general conditions would be more nearly balanced.

As a consequence of this physiological circumstance, the significance of the reports must be normalized with an exponentially decreasing value tied to the age of the tree.

Further, we are interested in *variation*, per se.

As we have said, volcanic eruptions occur in sudden outbursts that may be expected to produce sharp temporary reactions as well as maintain measurable average levels of dust in the upper atmosphere. Volcanic influence may cause sharply varying conditions—temperature drops followed by sharp rises, lots of rain followed by drought, and so on. While the averages look smooth, they are the products of many sharp changes.

When we look for the effects of volcanic activity in tree rings, we may look for a coefficient of variability, *as such*. We don't care about the direction of change, only the frequency and the degree.

For example, to test the proposition that tidal forces and tree-ring growth are related (presumably by way of volcanic dust veil) we persuaded our small computer to plot tidal forces *at perihelion* each year in the period from about A.D. 600 to 1950, then to plot tree-ring coefficient of variation for the same period, pointing out the degree of correlation.

Any bright high school senior can figure out a better and more informative test than this. If some kid does it, we'd like to have his results. The hooker is that he will have to provide his own computer, punching, and programming, as we did.

As it is, our dogged little machine plowed through perihelion tides using our program of Longman's formulas for tidal calculation. We processed the tree-ring data using a thirty-year sample that slid through the years in five-year increments, and the machine printed the stuff out in twenty-four hours of work.

Perihelion is always a time of high tidal stress and it is sometimes the point of *highest* tidal stress, but not always. Ideally, we'd figure out the tides for every day of the year and pick for comparison the date of highest tide in each year.

That would soak up 365 twenty-four-hour days of calculation on our little computer. (Not to knock this machine, a NOVA 820 with 8k of core. It is very fast, a miracle of logic and digital electronics, but the data and calculations we require are massive.) Time and budget limited us to something more modest.

Given that everything could be better, what did we get?

We found that tree-ring growth variation correlates with perihelion tidal variation with a level of significance at less than 1 percent. That is, p = less than 0.01.

To put it another way, if tidal forces and tree-ring growth are not significantly related, Dr. Browning will eat his hat in public.

The performance of the tree rings depends in part on where the tree is with respect to the equator. This is not just a matter of knowing how far the tree is from the volcanoes whose output controls the temperature and rainfall, since the effect is worldwide. (Note that when average temperature drops just 1° F in your home town, it's as if you had moved 300 miles higher in latitude—like moving from Indianapolis to north of Milwaukee.) We are concerned here with a critically important matter that will come up again in later discussions.

An effect of lowered temperature is the displacement of the planetary wind belts.

Consider Figure 5.4. The Earth turns from west to east quite rapidly. It's only one revolution a day—in terms of revolutions, only one-half the rate of the hour hand on a clock. However, the Earth is about 25,000 miles around at its equator, and any point on the

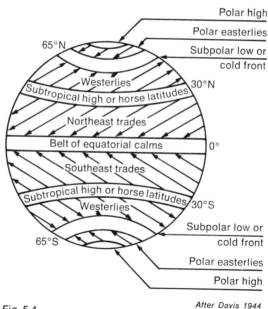

Fig. 5.4.   *After Davis 1944*

equator travels that whole distance in a single day. It sounds faster this way. As you move away from the equator, of course, you're not traveling quite so far; and if you get all the way to either pole and stand right on it, you will revolve gracefully just once a day. (Actually, the polar axis wobbles a bit, but you get the idea.)

Since Earth is moving so fast at the equator, it tends to spin the winds off toward the poles. It is these high-level winds trending gradually out to the poles which carry dust from equatorial volcanoes to the highest latitudes, causing the effect of cooling worldwide instead of keeping the effect local. When the winds reach the poles, they don't have anywhere else to go, so they swirl around and around, always west to east, until they lose energy. Even as they lose energy, though, they are being replenished by fresh breezes from the equator.

The swirl of air around the pole is called the "circumpolar vortex," as exotic-sounding a name as ever graced a page of technical discussion.

Americans will note that their television weathermen always indicate the general shape of the country's weather by sketching in storm fronts off the west coast. These fronts sweep across the country from west to east. Before them, tornadoes excite the populace. After them follow relief and clearing weather.

The cap of winds around the pole is not perfectly round. It is affected somewhat by the topography of the Earth below it, and its edge is scalloped irregularly so it trends up and down to the southeast and to the northeast. Further, the size of the cap changes. That is, the southern edge of the northern circumpolar vortex moves farther south during the winter, when it is colder. It stays in the south even during the summertime when the Earth's temperature is especially low.

This is very significant, because the southern edge of the northern circumpolar vortex is the line against which other wind systems bump when they move from one equator to the higher latitudes. This interface is the line of action, the place where the big storm systems develop, where the monsoons drop their rain, where the hurricanes and typhoons are formed. The patterns of the world weather system change significantly with changes in the Earth's temperature, which shift the circumpolar vortex.

These matters grow complex, but one effect deserves an attempt at explanation and understanding.

In general, it is hotter at the equator and cooler as one travels toward the poles to higher latitude. Temperature and precipitation are related. One might guess, then, that precipitation and latitude are related. Yes, but the relationship is not simple.

Moisture evaporates as well as precipitates. When precipitation is greater than evaporation in an area, the place tends to be humid. When evaporation is greater than precipitation, the area tends to be arid. The precipitation/evaporation ratio is to some extent a function of latitude. The graph that represents this relationship is anything but linear.

In 1953 Flohn published such a graph, and in 1971 Flint adapted and republished it, showing how things are "today" as compared with how things were in a glacial age.

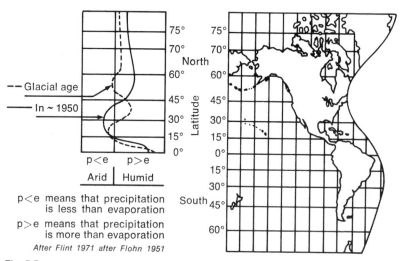

p<e means that precipitation is less than evaporation

p>e means that precipitation is more than evaporation

After Flint 1971 after Flohn 1951

Fig. 5.5.

This is worth some study.

The solid line indicates that today's humid zone extends from the equator to a high point at about five degrees of latitude, then drops off sharply. At about twelve degrees of latitude the arid zone begins, and it runs up to about forty degrees of latitude. There the line crosses the zero point again and another comparatively humid zone extends from about forty degrees of latitude to a peak at around sixty degrees, then tapers off gradually to the pole.

Look how sharply the glacial age curve differs. The lower latitude humid zone is less humid. The arid zone then extends only to about twenty-eight degrees of latitude. At that, the aridity is not nearly so pronounced. That is, the area is a lot wetter. The humidity increases very greatly all the way up to about forty-five degrees of latitude, where the curve crosses today's curve. After that, the glacial age curve drops very sharply and the zone from about forty-five degrees of latitude all the way to the pole is *much* drier.

This says, very simply, that when the temperature drops, the areas from the equator to about 15° get appreciably drier. The area from about 15° to about 45° gets very much wetter. The areas higher than 45° get much drier. Notice that at about 45° there is essentially no change in precipitation. Also, between about 40° and 45°, the drop in temperature counteracts the improved growing conditions produced by increased precipitation.

Tree rings are much affected by precipitation. In general, increased precipitation produces thicker tree rings. In this case, we are interested in thickness more than in variability.

To determine the effect of volcanoes on growing conditions, we went back through the literature and picked out the greatest volcanoes since A.D. 1500. In every case we assumed the volcano erupted at time zero. We then plotted tree-growth data at various latitudes in the years following the eruptions. In figures 5.6–5.10, the constant factor was the latitude of the trees. The volcanoes occurred at various latitudes and at various times. Since our tree-ring records were not complete for the same period of time in every location, the periods of years and the numbers of volcanoes differ somewhat from chart to chart.

The curve in each chart is an average representation of the effect of volcanic eruptions on tree growth over a period of years following the eruptions.

It appears from these graphs that Flohn's description of the precipitation/evaporation situation under different temperature conditions is confined by tree-ring response to volcanic activity. (Or make the other assumption, if you like, that the validity of Flohn's plot is established and the tree-ring response demonstrates that the temperature goes down after volcanic activity. Really, all we can safely declare

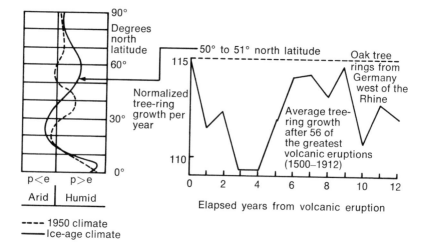

Fig. 5.6.

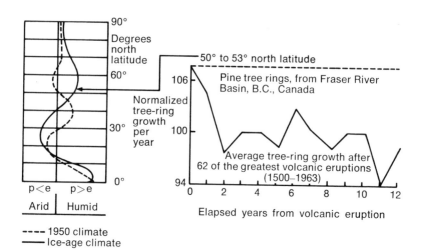

Fig. 5.7.

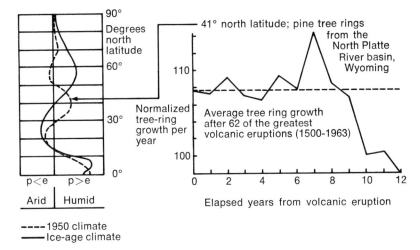

Fig. 5.8.

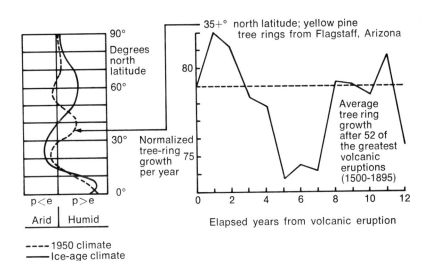

Fig. 5.9.

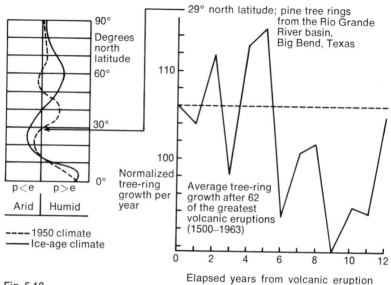

Fig. 5.10.

- - - - 1950 climate
——— Ice-age climate

p<e   p>e
Arid | Humid

29° north latitude; pine tree rings from the Rio Grande River basin, Big Bend, Texas

90°
Degrees north latitude
60°
110
30°
100
0°
Normalized tree-ring growth per year

Average tree-ring growth after 62 of the greatest volcanic eruptions (1500–1963)

0   2   4   6   8   10   12
Elapsed years from volcanic eruption

here is a correlation. You may choose for yourself what is cause and what effect.)

On the average, tree-ring growth above 45° north latitude is depressed for several years following a great volcanic eruption.

At 41° north growth is unaffected.

At 35° north and 29° north growth of tree rings is *increased* on the average for a few years after a great volcanic eruption.

This exercise gives us some basis on which to estimate the effects of decreasing or increasing temperature at particular latitudes. We may assume that one growing plant is rather like another in its response to changes in temperature, and thus, p/e ratio. That is, if the tree grows poorly, so does the wheat.

To proceed sensibly with the tree rings, we must here bring in a factor to which we have alluded briefly, but which we have not yet discussed—sunspots.

Sunspots occur in fairly regular cycles of number and size. The cycle is not rigid, as the Moon's eclipse node cycle is rigid. The mean peak-to-peak sunspot cycle is 11.375 years. Why? We haven't any real idea. It just happens to be so, as far as we can tell from the information astronomers are now able to offer.

More mysterious still is the *double* sunspot cycle, with a period of 22.75 years.

Suppose that you count sunspots during a peak year, then sit around feverishly awaiting the next peak year so you can count again. Chances are very great that one peak year will have a good many more spots than the other peak year has. Assume that the first peak you observed was a big one and the second peak, 11.37 years later (on the average), was a less impressive display with fewer spots.

If you are well equipped with instruments, you will also measure the magnetic polarity of each spot you see. Sunspots are apparently large magnets and they are always oriented in the same direction during a cycle.

For reasons that are obscure, the polarity shifts from cycle to cycle, along with the number of spots. If, in your first peak, the spots are all lined up with the positive poles toward the east, then all the spots that show up in the second peak will be lined up with their positive poles toward the west. Strange, what?

It is well known that during violent sunspot activity (usually in the big peaks, 22.75 years apart) electrical communications on Earth are strongly affected with "noise." This is no surprise when we think of wireless communications—we are accustomed to static in AM radio broadcasts and we know what a lot of racket thunderstorms cause—but it is surprising to learn that *cable* communications are also affected. Indeed, communications may be knocked out completely for hours at a time when a solar storm is active.

It is not widely realized that sunspot activity—again, chiefly the bigger peaks of the double cycle—affects precipitation on Earth strongly.

Reliable reports taken over many decades indicate that the levels of Lake Victoria Nyanza in Africa and Lake Huron in North America vary together. Further, the levels of both lakes vary with the double sunspot cycle.

Of course, this shows up in the tree-ring records. Both the 11.37- and the 22.75-year cycles are detectable. The effects of sunspots must be considered in relation to tidal cycles, and no calculation of future effects may safely be based on tidal cycles alone. Sunspots must be taken into account. No, we don't really know why (though there are some theories), but we know there are distinct correlations between sunspots and rainfall, and it would be arbitrary to leave those correlations out of our calculations just because we don't know what is behind them.

The sunspot cycle question has been pressed most vigorously by C. G. Abbot, whose 1913 publication we have already quoted here. Dr. Abbot was a fan of cycles in general and a particularly determined researcher into the effects of sunspots. Here is another example of Abbot's work.

He plotted the decrease in solar radiation arriving at the Earth after volcanoes (curve A).

He combined this transmission coefficient of the atmosphere with sunspots (curve B) to produce a figure of merit (curve C). This

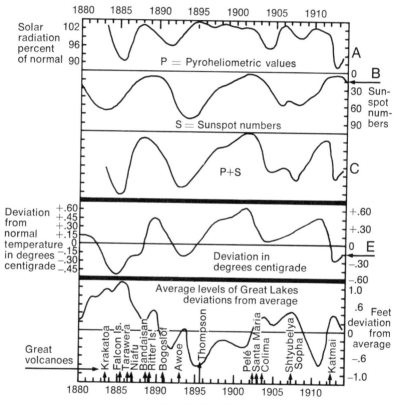

After Abbot and Fowle 1913; Tannehill 1947; Great Lake depth figures from NOAA, U. S. Dept. of Commerce

Fig. 5.11.

curve has a high degree of correlation with the mean annual temperature of the world (curve E.)

We have added the bottom curve showing variations in the level of the Great Lakes.

As recently as 1972, Rob Roosen told us of correspondence in which Dr. Abbot complained that few people were taking his cyclic work seriously, and only Krick, out in Denver, was sensibly using his data. Rob says that Abbot was even then undertaking a major task to demonstrate more clearly the usefulness of the cyclic observations. (Abbot worked mainly on solar power in his last ten years. His weather work had been totally "discredited.")

Dr. Abbot was over one hundred years old when he died in December 1973. In the summer of 1973 a crater of the moon was named for him. He is the only man ever so honored within his lifetime —at the suggestion of the *Russians.*

Let us indicate here a number of the significant periods we have observed in the tree-ring data and point out some of the components that may produce the composite longer periods.

Recall yet again the neighbor's pets. The regularly spaced tasks produced peaks of activity and lulls, no matter how the various tasks were phased. Looking at our tree-ring periods, let us break them down into similar task periods.

Do all these fanciful things really show up? Yes. You have to dig for them sometimes, but they come through the records sharply if you are tuned for them.

LaMarche published a fine study in 1974 which demonstrated at least one of the cycles—800 years—very clearly.

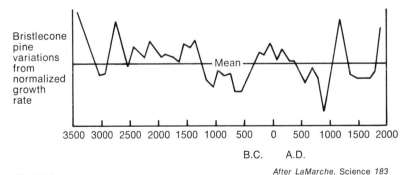

Fig. 5.12.

After LaMarche, Science *183*

Table 5.1.

Cycle detected	Components
1 year	Perihelion
4.425 years	Perigee precession ½ cycle
9.3 years	½ eclipse node cycle
11.77 years	Sunspot cycle
22.75 years	Double sunspot cycle
45 years	Two double sunspot cycles, almost 10 x 4.425 years, almost 5 x 9.3 years, almost 45 x 1 year
177 years	20 x 8.85 years (perigee, perihelion lineup) almost 20 x 4.425 years, almost 19 eclipse node semicycles, almost 177 rounds of perihelion

Note: the 11.375-year sunspot cycle fits into 177 years
15.5604 times, putting it almost exactly out of phase.
However:

354 years	All elements of 177-year cycle present, plus sunspots in phase
801 years	4 x 177 years plus 93 years 801 years (sunspots out of phase) 801 rounds of perihelion

Note: 93 years = 10 x 9.3 years
92.925 years = 21 x 4.425 years

1,739 years	393 x 4.425 years, almost 187 x 9.3 years, almost 21,508.5 synodic months, 1,739 perihelions Sunspots in phase

Notice that Figure 5.12 presents tree-ring growth departure from mean growth by century for about five and a half millennia. The graph starts at a peak around 3600 B.C.

Eight hundred years later, about 2800 B.C. a sharp change is under way.

In 2000 B.C. the change is visible, but less pronounced on this hundred-year average unit graph.

Twelve hundred B.C. marks another sharp change.

Four hundred B.C. is on the slope of another sharp change.

A.D. 400 falls on the knee of a change.

A.D. 1200 marks a great peak.

In the 1970s we appear to be climbing toward another significant peak around A.D. 2000.

Notice an anomalous sharp change around A.D. 650 which led to a very sharp drop. This was an interesting time, which we'll talk about later.

As we have suggested, a plot averaging activity over a century at a time effectively masks shorter periods. This particular type of plot does not serve our needs especially well, though it serves others just fine and no criticism is intended. Even so, the peaks and valleys are obvious and one can get from this a feel for the major cyclic activity with which we are concerned.

Second (if you can remember all the way back to when we started talking about tree rings "first"), ice cores provide a record of history.

We have said that the layers of snow-turned-to-ice can be dated and analyzed to reveal the temperature at which they were formed. The mechanism of the analysis is to measure the amount of the isotope 0^{18} present, compared with the amount of the more common isotope of oxygen, 0^{16}. The ratios of the isotopes vary reliably with the temperature at which ice forms. Thus, the temperature information we seek.

Date is determined by counting the layers from some known point.

The amount of precipitation may be determined by measuring the thickness of each layer.

Like everything else, this sounds simpler than it is. For one thing, as the ice gets thicker and thicker with the addition of layers, the pressure on the lower layers increases. By the time thousands of feet of ice have accumulated, the bottom levels are being squeezed fiercely and they warp and spread out as you'd expect them to. This means that the readings change as you go deeper in the ice—further back in time.

Researchers have looked into this and have learned to apply corrective factors to the older levels just as we apply compensating factors for the growth of rings through the life of a tree. (As a matter of fact, the last few years have seen a considerable revolution in the use of radioactive carbon dating—a technique that has been very widely publicized. The assumption was that plants have always taken up radioactive carbon at a steady rate. In fact, it has apparently *not* been steady over the millennia. In recent times, the rate has been steady, but as you go back, it seems to have changed. Many articles dated before this realization were actually a couple of hundred years older than the carbon dating indicated. The discovery has relieved the minds of a lot of archeologists who were completely baffled by the discrepancy between what they believed in their hearts and what the carbon readings told them. By the time you go back five thousand years, you may be off a thousand years. The curve is not simple, but complex. Now that the correction factors are known, we can use carbon dating with restored assurance. Same thing with tree rings and ice cores.)

Figure 5.13 shows the temperature record indicated over the past eight hundred years by ice-core studies made at Camp Century in Greenland. This chart can't stand by itself in telling us what has been going on in the world's climate, but it points up some changes in which we are interested, especially the period of the thirties to the fifties of this century, in which the temperature was extraordinarily high, higher than anything else for several hundred years.

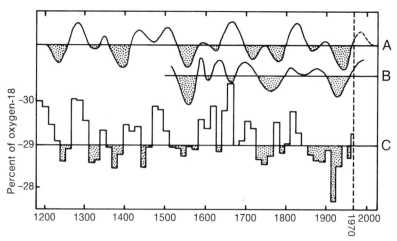

Curve C is the percent of oxygen-18 in Camp Century, Greenland, ice.

Curve A was synthesized from two harmonics—181 and 78 years.

From Dansgaard et al, 1971

Curve B is the derivative of the curve of the envelope of highest tides at 30° north latitude.

Calculated by Harrington, Naval Observatory, 1974

Fig. 5.13.

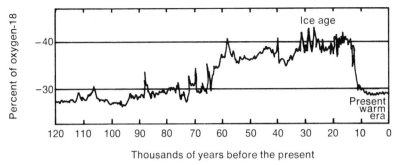

Fig. 5.14.

After Dansgaard et al 1971

Figure 5.14 shows a much longer period of time plotted from the Camp Century core. This runs back some 120,000 years, well into the period that requires the compensation factors represented in this graph.

The earlier references to the work of Gow and Williamson involved a similar core from Antarctica with the layers of volcanic ash that appeared in correlation with dramatic changes in the temperature indicators.

Third, cores taken from the bottoms of the oceans and lakes show off the *varves,* the layers of sediment which are deposited year by year. These layers can be correlated well with the ice-core layers.

In the sea, of course, one finds all sorts of biological materials, microscopic organisms with growth characteristics which can be readily correlated with temperature, nutrient supply, etc. Tektites and microtektites are also found in the varve samples. The ocean bottom is teeming with data.

Lake bottoms turn out to be especially interesting. When a lake has no outlet, everything that falls into it stays there and collects nicely at the bottom. Lakes thus build up layered bottoms that preserve pollens from ancient times, letting us determine what sort of vegetation grew when around those lakes. It is interesting to note that man is just barely detectable in the pollen count of European lakes of Neolithic times. That's when man undertook slash-and-burn programs to clear the land so he could plant crops. The pollen from those cultivated crops and related weeds shows up in the lakes just enough to indicate what was occurring. The ratio of cultivated to wild plants shifted once again to mark the visitation of the great plagues in Europe in the thirteenth century, which killed such a large percentage of the human population that vast acreages of cultivated land reverted to the wild for want of care.

A depression (perhaps a meteoritic crater) on St. Paul Island in the Bering Sea north of the Aleutians has been a collector of pollen and various other things (the place is used as a reindeer corral in recent times) for many thousands of years.

At the moment, the island is isolated, well out in the sea, well away from any reliable supply of pollen. However, the varves in that depression indicate the presence in times past of many kinds of vegetation, notably a birch forest. Apparently, when the sea level drops and the Bering land bridge is above the surface, St. Paul is not an island, but a hill, surrounded by growing things.

Last, let us mention geological strata records. Really, these tend to be sedimentary layers from the bottoms of seas or lakes which have been around for so very long that they have turned to stone instead of mud.

Anderson and his associates[7] reported on drillings in the 1,500-foot-thick laminated siltstone evaporite of the Castile and Salado formation in southeastern New Mexico, overlapping into Texas. The layered evaporite precipitated to the bottom of a 2,000-foot-deep bay some hundred miles across. The bay had a narrow mouth connection to the ocean and that mouth had a shallow sill that helped to isolate the bay. The layers give us a tree-ring-like record, season by season, of weather spanning some 250,000 years—200 million years ago (early Permian).

The core from this ancient bay shows us the same periodic cycles that appear in the very much more recent ice cores and tree rings. The $\sim$ 350-year cycle, for example, is present in the record.

Unlike ice cores, however, this record shows *no ice ages* in all that time. The evaporite was formed in what we would now regard as a period of quite constant climate—as, indeed, we know the Permian to have been. This climate is much like our immediate past and what

7. R. Y. Anderson, W. E. Dean, Jr., D. W. Kirkland, and H. I. Snider, "Permian Castile Varved Evaporite Sequence, West Texas and New Mexico," *Geological Society of America Bulletin*, vol. 83, 1972.

appears to be our immediate future. For 30,000 years, this peaceful sort of climate should prevail again.

Our descendants, not more unlike us than we are unlike Cro-Magnon man, may well bask in the more quiet periods without the stimulation of changing temperatures and precipitation brought to them by great volcanic activity.

Paradise, perhaps.

With equal or greater probability, hell on Earth.

6

Human beings notice dramatic events like bolide strikes, earthquakes, and volcanoes, but it is not intuitively obvious to the casual observer that changes that fail either to melt the ice caps or to bring on new ice ages can be significant. Since we are concerned with human affairs, it is worthwhile to make some observations here about people.

In *The Achieving Society*,[1] a superb work on human behavior, David C. McClelland made an excellent case for the proposition that the performance of human beings is largely a function of the average temperature at which they and their cultural forebears have spent their lives.

McClelland concludes that the personal need for achievement (a characteristic he defines and discusses at length) is greatest in countries whose center of population is on the line marking 50° F average temperature. The performance of the whole society reflects the individual need to achieve—as contrasted with the need to gain power or the need of affiliation, or whatever.

McClelland was unable to account satisfactorily for the comparatively rapid changes that occur in the performance of whole societies. He points out properly that genetics among human beings are unlikely to be altered significantly in a few decades. Less properly, he

---

1. D. C. McClelland, *The Achieving Society*. New York: Van Nostrand, 1961.

suggests that climate can't change rapidly enough to change attitudes in just a few decades. Yet the performance changes that actually occur are exactly what you'd expect if a whole society were picked up and transported hundreds of miles north or south to cooler or warmer climes.

Even discounting such events as instant deep freezing, climatic changes have a profound effect on human affairs.

Let us begin with long-term effects. We have quite a clear view of human modification which is attributed to known climatic change. Recall an event with which we are all somewhat distantly familiar, the crossing of the land bridge from Siberia. The greater number of inhabitants of the Americas who were here when Columbus came were evidently descendants of the people who walked here from the Old World many thousand years ago when more of the world's water was tied up in ice and the level of the sea was lower. At that time (and at other times in the past) the shallow Bering Sea retreated and left a bridge of land between Siberia and Alaska. People could cross that bridge and apparently they did so. Mongol peoples left Asia in waves and came to the Americas, spreading clear down to Tierra del Fuego in the course of time. They were the ancestors of the New World Indians.

Most of us imagine hardy little bands of travelers working their way diligently across a narrow isthmus from west to east. We see them trooping in single file while their dogs trot inquisitively around them, sometimes rushing down the embankment to the sea to investigate an interesting object that the waves have washed up.

The travelers glance frequently toward the distant horizon, where the volcanic peaks of ancient Alaska rise to mark the end of their unnerving journey, of their many days of exposure to the elements and to enemy attack in a constricted and unfamiliar space.

They breathe more easily as they leave the narrow causeway and move onto the strange continent where they and their descendants will struggle with dangers for thousands of years.

A bold and touching picture of these hardy pioneers, is it not?

Inaccurate, too.

At times, apparently, the Bering Strait land bridge was not merely hundreds of yards across, not just a mile or two across, but *thirteen hundred miles* across.

That's not just a little, narrow neck of land; that's a continental mass. People could have crossed the Bering Strait land bridge over several generations, moving a bit at a time, building settlements, raising families, having wars. Large tribes may have migrated gradually eastward over the centuries and diffused into North America without ever seeing the ocean or even knowing, except in legend, that there was such a thing as a great expanse of salt water.

Those waters rose and fell rather slowly, but the change clearly affected man's lot. It is interesting to note that migration to the New World has apparently occurred over a period of at least 27,000 years, perhaps appreciably longer.

Mongoloid peoples who left Asia 25,000 years ago (or so) and moved down through the Americas were people whose physical traits had been established in a warm area. Notably, they had no epicanthic fold, the fatty layer in the eyelid which is so noticeable to Caucasians. These were obviously not people whose ancestors trapped for fur in the snowy wastes of the polar regions or speared seals from floating ice. They were fairly dark-skinned people whose ancestors had lived in warm country.

By the time the Mongol Athabaskan peoples came to the Americas (presumably by water, for the land bridge was gone) about three thousand years ago, the epicanthic fold was already developing and the Navajo of today, descended from those people, is likely to have a touch of the Oriental in his appearance.

When the Eskimo reached the New World only about a thousand years ago, the epicanthic fold was developed to an extraordinary degree.

The Eskimo's eye is designed for cold weather. The heavy layer of fat insulates him so that his eyeball won't freeze when the temperature becomes nippy while he's sleeping. The fold is so heavy that his eye usually opens to but a slit, protecting him from much of the glare of the Sun on the ice and snow. His eye didn't develop this excellent cold-weather configuration in a tropical rain forest. His people spent a long time in chilly places, and everybody whose eyeballs froze was eliminated from the gene pool that makes today's Eskimo successful.

As we have indicated, the Mongols are newcomers to the cold country and they have developed the useful "slanted" eye to handle the change in their surroundings. The oldtimers in the north are people like the Ainu, the "Hairy People" of Japan. "Hairy" is almost an understatement. These folks have enough body hair to weave a set of long johns.

The Ainu are a Caucasoid or proto-Caucasoid people who were once widely distributed through northern Asia. At some point the warm-weather people pushed into their territory and took it over. These invading Mongols not only developed epicanthic eye folds, they developed an aversion to Ainus and warred with them constantly for thousands of years. The issue was not finally settled until the Ainu had been shoved into a corner of northern Japan and were effectively contained. The Ainu disapproved of this containment and as recently as the mid-1800s launched a major effort to break out.

They lost. The Mongol people who took Japan from them beat them back with ferocity. "Pacification" was achieved only when it became quite clear to the Ainu that they faced annihilation if they continued to annoy their conquerors. Today they are an anthropologist's delight, but an embarrassment to the Mongoloid Japanese, who feel a bit guilty about people to whom they have heretofore felt morally superior. A sort of "Bureau of Ainu Affairs" brings the blessings of bureaucracy to the Hairy People and presides over the disintegration of their ancient ways. Within recent years there has been civil rights agitation among the Ainu, but there's not a whole lot that a trifling racial minority without a grasp of modern technology can do to compete with an aggressive and energetic majority.

The depth of Japanese/Ainu disaffection astonishes both Yankees and the Japanese when it is revealed in small ways. For example, three college girls in California were making much of a baby boy when one commented that it was strange to think that the little cutie would, in a few short years, turn into a big, hairy man. The Japanese girl in the group was almost nauseated at the thought and said so. One of the other girls laughed and said that men are uninteresting until they have heavy hair on their chests. This was too much for our Japanese girl, and she stamped around indignantly, going, "Ugh!" One got the impression of a deep-seated cultural aversion to hairy people in her background. This girl is a third-generation American!

It took a while, but in Japan the warm-country people adapted to the cold weather and beat the stuffing out of the resident Caucasoids. In North America, the opposite process moved a bit faster when the Caucasoids came in already accustomed to rough weather and beat the stuffing out of the resident Mongoloids.

This course of events can be traced with remarkable clarity, including the physical adaptation of a people to a major environmental change. There are many such physical developments that we can trace as we learn more.

Assuming that we all started from Adam and Eve—or even if we didn't—we are one species. Human races are not so unlike that they cannot intermarry and produce succeeding generations of fertile offspring. The racial differences developed after some time of greater similarity. Why?

Here's a hotly disputed view of the matter which fits comfortably with the proposition that climatic changes govern human affairs.

There are evolutionary pressures that change the heredity of people. These outstanding genetic pressures are produced by:

1. Sunlight.
2. Latitude.
3. Violent conflict.
4. Genetic drift.

All these take:

5. Time.

Obviously, there are other, lesser pressures. It is not our purpose here to survey the whole field of human genetics, but to point to some areas that seem to deserve additional research.

Consider these pressures one at a time.

1. Sunlight causes skin cancer. The relationship of skin cancer to solar radiation has been demonstrated too many times to need much elaboration here. One example may suffice: It has been noted that women in England tend to develop skin cancer on their left sides, preferentially, and men develop skin cancer on the right side.

In the United States, the opposite is true: women on the right, men on the left.

An English lecturer once presented this with dramatic flair by working with a marking pen on store mannequins. As he recited the statistics about the locations of cancers, he marked the spots on the dummies. Before one's very eyes the patterns became obvious.

As the lecture proceeded, the parts of the dummies which would normally be clothed remained comparatively clear while the unclothed areas gradually grew mottled. The sleeves and necklines were outlined. The positions of the cuffs and the seams were obvious.

And, strangely, the right and left tendencies of England and America showed.

The only explanation that seems plausible is that the English drive on the left side of the road with the driver on the right. The Yankees drive on the right side of the road with the driver on the left side of the car. In both countries, the men do most of the driving and the women spend more time in the front passenger seat. American men get more exposure to sunlight through car windows to their left, women to the right. In England, the reverse is true.

Really? Is this the reason for the difference? All we can do is report the news.

Skin cancer incidence is inversely proportional to the amount of melanin pigment in the skin. Whites are highly susceptible to skin cancer. Blacks are not. Everybody in between has a skin cancer probability that diminishes with the amount of pigment in his skin.

Skin cancer is deadly, rather like having your eyeballs freeze in an unguarded moment. People who are susceptible to skin cancer gradually vanished from populations in countries where there's a lot of hard sunlight.

Dark skin is a great advantage for people who live in equatorial regions. Thick, matted hair on top of the head is a big help, too.

It would seem worthwhile for everybody to be very black with bushy hair. However.

2. Latitude makes a difference. Climate at the equator tends to be fairly consistent. The daily and seasonal excursions in temperature are not very great and vegetation tends to be heavy year round.

An individual in such circumstances has little trouble in getting enough vitamins. Some vitamins, like C and D_1, are not stored in the body, and it is necessary to have constant fresh supplies of them.

At high latitudes the temperature varies a great deal daily and seasonally. A forager in Poland, for example, will become discouraged as he searches the snowbound countryside for fresh fruit in the wintertime. He can obtain vitamin C from potatoes that he has stashed away and from other vegetables that keep well, but the vitamin D is another matter.

Actually, the body will produce vitamin D_1 itself if a person is exposed directly to sunlight. The effect of sunlight on production of vitamin D_1 is inversely proportional to the amount of melanin in the skin. That is, if vitamin D_1 is not available in fresh food in Warsaw, it's no help to be black. For one thing, the sun doesn't shine all that often or all that hard in the high latitudes, so the danger of skin cancer is reduced. At the same time, there is a great premium on the ability to produce vitamin D_1 in the presence of what little sunshine there is. People who can't do it—in general, people with dark skins—have been selected *against* in the high latitudes.

This vitamin D_1 business is a sharper differentiator than we have realized until recently, when a lot of good scientific work has been done indicating that adult Caucasoids can drink milk, by and large, while adults of other races can't, by and large.

Milk is loaded with vitamin D_1, as the National Dairy Council has been telling us for years. People who need D_1 in high latitudes have a good thing going if they can drink milk. For those who can stand it, it beats eating vegetables that have been in a root cellar for months. (Of course, you still need your potato skins for vitamin C, but . . .)

This subject came up when people began to notice that, while Danes and Swedes in Wisconsin guzzled milk happily, the black population of Chicago was rather less enthusiastic about it. Similarly, it turned out, 85 percent of the people in Thailand over five years old can't drink milk. The Navajos and others in New Mexico and Arizona regard milk with some hostility.

Schoolteachers and nutritionists inflamed with zeal to do good to the Ainu—er—, the Indians—have for many years fought a grim battle with Indian kids who obstinately refuse to drink the good milk provided to them by the government. Every Anglo knows that milk is the stuff upon which strong bones and teeth thrive. Every Navajo kid knows that milk is poisonous. The stuff gives him diarrhea or makes him throw up.

Indian kids in the Southwest have received copious quantities of milk under the Johnson-O'Malley Act, presumably on the theory that what's good for Anglo senators is bound to be good for Indians. The J-O kids, as they are known in the trade, have faced torment as the milk is pumped to them.

Evidence close to home confirms this subjective effect. Dr. Browning is one-sixteenth Cherokee. His sixteenth includes the part that can't stand milk.

Anybody who can get his hands on the Life series *Foods of the World* can do the same statistical experiment we did when rumors of this came to us a few years ago. We went through the books, which are broken down by region, and figured out what percentage of recipes in each area included the use of dairy products. The percentages vary greatly. You may guess that the Danes are big on cooking with milk. You may guess that the Japanese are not.

More convincingly, dig through the U.N. figures and other references that provide racial breakdowns on the countries of the world. It's a lot of work, but we'd already done it for other purposes, so we could run this experiment with no strain.

You will discover that the percentage of recipes using dairy products in any given country corresponds *very* closely with the percentage of Caucasian genes in the population of that country.

We were interested largely because we were aware that the dairy industry in the United States is on hard times. Some of the dairy companies whose names we have known for many decades are sliding out of the fresh milk business. Many dairy herds have been served up as hamburger. It seems a shame when large parts of the world are desperately in need of the protein that milk could provide. We wondered why milk isn't in demand in Africa and China and Japan. We wondered why the Navajos mix the powdered milk they receive into whitewash and paint sheds with it.

The problem seems to be lactose, the natural sugar occurring in all milk (including human milk). Babies can drink the stuff because their bodies produce an enzyme that breaks the lactose into other forms of sugar that the body can use. Growing children tend to lose their ability to produce this enzyme and gradually they become allergic to lactose or otherwise react unpleasantly. You can imagine what would happen if you put large amounts of highly nutritious food into your intestine when you are unable to digest it, but the bacteria that normally reside there can. You become cultured, as it were. Apparently, Caucasians at high latitudes, needing vitamin D_1, were selected in favor of a tolerance for lactose. Any other race can digest any dairy product (such as cheese) which has had the lactose destroyed by fermentation.

When we showed our silly cookbook study to friends who are in the business of producing enzymes to do jobs like converting lactose to something else, we were treated with some disdain. Our information didn't sound very scientific, so Randy Davis went off and dug up good, solid studies that we had not seen. He reported to us in some excitement that there was *real* evidence we weren't just pranksters. With a perfectly straight face Randy told us that "cats, rats, bats, ferrets, weasels, polar bears, and adult Caucasian human beings retain their infantile tolerance of lactose." Some of us adult Caucasians and weasels find that funny.

We carried the matter further, suggesting to a major milk producing company that they might open worthwhile new markets by using some techniques we worked out for recovery of enzymes that convert the lactose for non-Caucasians. They wondered about the validity of the basic idea, pointing out that a study of U.S. Negroes showed no such sharp distinction in lactose tolerance between them and Caucasians.

We inquired if the study had made allowance for the fact that roughly 30 percent of the gene pool among U.S. Negroes is Caucasian. Well, uh . . . no.

We suggested that they send a research group up to Oakland to work with the Black Panthers in their free-breakfast-for-the-underprivileged program. It turned out that the milk company thought of research as something you do in a white coat in a laboratory, not something that takes you to the middle of the loudest social hassle you can find. The matter seems to have died.

It seems unlikely that treated milk can be made useful, as such, to people whose traditions have taught them that milk is bad news. The stuff must be made socially as well as technically acceptable. We tried working up a deal between the Mormon church and the organization of witch doctors in Africa, hoping that their mutual desire to save souls and do-the-right-thing could help to save lives as well. Between them, they might have financed the project and made powdered, converted milk socially desirable. It was hard to get anybody to take us seriously.

The idea is not universally accepted that intolerance of milk is a genetic effect. People who are determined to make the Indian kids drink milk are convinced that their resistance is cultural.

That idea was tested in a Japanese study. A number of Japanese volunteers were fed milk on a regular basis for two years so that they would grow used to the stuff despite their traditional cultural avoidance of it.

After two years, the most notable effect among the volunteers was their great relief that the experiment had come to an end. While they had learned to gag down milk on a sort of "Banzai!" basis, they had been unable to persuade their digestive systems to view the matter philosophically. No, the problem doesn't seem to be cultural.

By the way, we do not mean to suggest that the 30 percent of Caucasian genes in the U.S. black population is exclusively the product of interracial activity that occurred after the blacks were brought reluctantly to the New World. The fact is that color, per se, has nothing to do with "race."

The Australian Bushmen are considered by some anthropologists to be Caucasian or proto-Caucasian, though they are very dark. Similarly, the Watusi of central Africa are very black, but Caucasian by most other standards. They are mixed—a "Nilotic" people in whom black pigmentation is dominant genetically (and a good thing, too, if they are going to live where they do).

The Watusi are an interesting anomaly in an interesting area. They are very tall people. Indeed, tallness is an enormously important matter among them culturally, and the tallest of them choose to marry tall partners, thus selecting for height. The very tallest Watusi is the king.

The ability to jump high is also valued greatly, and the kids practice hard to be good jumpers when they grow up. Indeed, the test for governmental office has traditionally included high jumps. Low jumpers and little guys simply can't pass this civil service exam. The selection for height is reinforced over and over.

That great, strong, long bone structure is aided by the ability to drink milk—a Caucasian characteristic that is retained in this cultural emphasis on height.

The Watusi in centuries past moved from higher latitudes to central Africa and established a position of dominance over the truly Negroid Hutu people. The Hutu herded cattle for the Watusi, but did not drink the milk of those cattle as the Watusi do. In the same general part of the world, the half-Caucasian Masai herd cattle and drink milk mixed half and half with blood.

It may make for a certain uneasiness among very tall, high-jumping black U.S. basketball players to contemplate the proposition that they are Caucasians along with us weasels and polar bears.

Other Caucasians who acquired their genes in high latitudes include the now-brown Hamites, the light-brown Semites and Mediterranean subraces, and the brunette Alpine people along with the Ainu and the blond, blue-eyed Nordics.

A large group of Caucasians migrated from the area of the Caucasus clear down into southern India, where they dominated the local folk, non-Caucasian Dravidians. These invaders were the Aryans, regarded in later times as virtually the prototypes of Caucasians.

They have presented India with a rather peculiar situation ever since they arrived there. The Hindu religion (whose sacred writings are the Rig-Vedas, which appeared in India, written in Sanskrit, along with these Aryans sometime between 1500 and 1200 B.C.) is very strong on the proposition that cows are not to be eaten. Cows may be worked. They may be milked. The milk may be consumed, but the cow may not be consumed.

This is great for people who can drink milk and use dairy products in other ways, but it's hard on people who cannot tolerate lactose. India is surrounded by people who are not Caucasian and cannot drink milk. India has never been successful in exporting the Hindu religion to its neighbors except in the form of Indians (example: Guyana in South America, which used to be British Guiana). Buddhism started in India and successfully moved out among the Mongol people all around, but only the Aryans seem to find solace in Hinduism (barring, to be sure, certain exceptions). Hinduism has not swept the world from its home base. One wonders if the inability of the most likely converts to stay alive while following the rules has anything to do with the problem.

It appears that the latitude at which people develop their racial characteristics has a major effect on their color, hair type, and diet.

What else?

Shape of the nose, for one thing. If you live at very high latitudes where it gets 'way down below zero often, you are in danger of freezing your lungs as you take a deep, invigorating breath. People who freeze their lungs during the winter tend not to be among the living come springtime.

There is selection at high latitudes in favor of those who are able to heat up frigid air somehow before it freezes their lungs. The mechanism for this is a "radiator," a heat exchanger called the "internal meatus," which is found in the noses of people whose ancestors survived thousands of long, cold winters. The internal meatus not only takes the chill off of air headed for the lungs, but provides a big nose with a prominent bridge to its wearer.

The internal meatus does something else. It provides a great, big pollen catcher to its wearer, and if he is allergic to pollen the whole thing swells up and prevents his breathing through his nose.

The variety of pollen is limited at high latitudes. The variety of pollen is very great in low latitudes where the vegetation is rich.

Thus, there is selection against the internal meatus with a big nose in equatorial areas.

Caucasians tend to have big, sharp noses with pinched nostrils. Negroid peoples tend to have flat noses with large nostrils suited to the equatorial climate.

3. Violent conflict is an obvious genetic selector that probably selects for intelligence, tool-handling, and a large, healthy body. No visible effect of latitude in this. No matter where, people who are too weak, too dumb, too slow, tend to be eliminated from the competition before they can have children who are like them.

The ability to handle tools and intelligence are closely related to other human characteristics. One view of human evolution may be summarized this way:

Some early humanoid found that he could do things with a tool that he held in his hand. He found it very inconvenient to walk on all fours with the tool in his hand, so he tended to straighten up and walk on his hind legs. The possession of the tool gave him a selective advantage. Over many generations, the tool-handlers did better than the non-tool-handlers. Thus, there was selection for tool-handlers and hind-leg walkers.

The use of tools was related to development of the brain, and there was selection for intelligence. At the same time, the birth channel grew more restricted in size, owing to the erect posture. With the brain growing and increasing the size of the cranium, this created conflicting pressures. Yet, the advantage of intelligence outweighed other problems.

There was selection for babies that were born prematurely, by ordinary mammalian standards. Babies with heads small enough to make it through the birth channel survived more often than others. In addition, the head of the infant was soft, so it could be deformed during birth without permanent damage. This has led to the present status in which human beings have the shortest gestation period of any mammal of their size, except for marsupials, which operate under a different set of pressures, with different solutions to the problems of premature birth.

Since human babies are born so early in their development, perhaps nine months sooner than would be the case without this evolutionary change, the primitively formed child is exposed to the conscious training efforts of his parents at a time when he is especially receptive to training. Human babies are taught different attitudes by their parents very early in life. The lessons stick.

It is worthy of note that a study in Mexico indicated that intelligence is strongly affected by the availability of animal protein in the diet. Kids who live on protein provided by beans only may have an IQ disadvantage of as much as fifteen points. The ability to catch or raise and eat animals is important. The ability to drink milk also allows for intake of animal protein. Where animal protein is plentiful, intelligence is aided.

This genetic emphasis on intelligence is not limited to man. Some suggest that the opossum is a good model for us to admire in the way of getting along in the world without adaptation. They point out that this marsupial has been among us unaltered for fifty million years.

Well, not quite. While the opossum looks just about as he did ever so long ago, there has been a perceptible change in his makeup. His brain has steadily increased in size. He holds the same ecological niche he always did, but he has to get smarter and smarter just to stay where he is. *All people* are smarter than anything else around. They must keep growing smarter to hold their place.

4. Genetic drift, or rapid random evolution, is a chance occurrence in an isolated small community. There are remote communities in which it is almost a novelty to be born without six fingers on each hand. In Africa, the Capoid people became isolated for a long time at the high southern latitudes. The adults had hereditarily modified external sex organs; the fat was all deposited on the rump; the skin was, compared with the rest of mankind, excessively wrinkled. Their color matched the latitude—yellow, not black or white. Characteristics like this are developed in remote areas, then added to the more general population when the isolation is broken. The African Bushmen are largely Capoid. The Hottentots are part Capoid, part Negroid.

5. And, of course, all this takes time, many, many generations. The "original" races must have developed over a very long period of climatic stability which made possible the sharply defined genetic differences.

Notice that human beings are estimated to have about thirty thousand genes with which to transfer their characteristics to succeeding generations. Of those, only about two thousand are factors that may account for differences among the races.

The structure of the chromosomes upon which the genes occur

and the general chemical structure of human beings does change with time, even in a homogeneous population that stays in one place for a long time.

It has become possible in recent years to plot these changes in populations over long periods of time. The method is useful, since it allows us to determine a number of handy things about ancient peoples who left no specific data about themselves. Now we can examine mummies, accidentally or deliberately preserved (there are a lot of frozen and dried and potted people around), and figure out when those people lived in the line of their race. From these samples of what may best be described as "jerky," the dry meat of somebody who doesn't care any more, we can tell where they came from, who their ancestors were. We have discovered that a body in an ancient tomb is that of a newcomer to that area—perhaps a conqueror who ruled people to whom he was not related.

We have determined that a mummy in a twelfth-dynasty Egyptian tomb is a ringer—somebody stuck in the old tomb hundreds of years later than the era in which the structure was built.

We have discovered that the earliest known inhabitants of Scotland had genetic and chemical structures just like those of Egyptians of the same era. This leads to profound suspicion that those early Scots were black refugees or colonists from the land of the Nile. (This doesn't seem to fit the image of the kilt and the tam, but one is prepared to believe that the ancestors of haggis-eaters were invited to leave Egypt with their ideas of cuisine.)

Less exotic techniques than fossil pathology indicate human developmental history. For example, people who have spent a very long time in equatorial regions are doubly, triply, and quadruply protected against skin cancer by the overlay of many mechanisms that assure dark pigmentation. Not just one gene is involved in the production of pigment, but many. This is demonstrated by the observation that the children of parents who are white Caucasian and pure Negroid, respectively, tend to be colored some shade in between white and black, but virtually never "white." The many genes controlling color dominate and assure dark pigmentation in the child.

On the other hand, the children of pure Australoid and white Caucasian parents will be either "white" or "black." This suggests that a single gene may control the pigmentation. It suggests further that the Australoids haven't been in the hot country very long (we know they went to Australia about twenty thousand years ago) and they haven't developed the protection in depth against skin cancer which the equatorial people have developed over perhaps two million years.

The multiple-gene mechanism protects against almost all accidents of light color, but not quite all. A famous photograph of some decades ago shows an albino African posing with his very black tribe. He is conspicuous in the crowd. He had every disadvantage that light skin could offer. Selection against albinism at that latitude is very powerful.

There is reason to think that latitude not only affects physical racial differences, but cultural differences among people as well.

Again, our purpose here is to report the news, not to make value judgments. Differences do exist. One may decide for himself whether to be a pure relativist in this matter or to be a partisan of some particular viewpoint. Our own bias here is to look upon traits that contribute to survival as "good" and traits that don't contribute to survival as "bad." We must assume that all people, like opossums, have to get smarter all the time to keep their niche. The survivors in any area and in any race are those who were smart enough to do something "appropriate" whenever a question or a problem arose. "Appropriateness" varies with latitude.

For example: the notion of "properly modest" clothing seems to depend on the average temperature and variability of the weather. It is the high-latitude people who traditionally develop strong moral notions about keeping oneself covered. The U.S. missionaries who went to Hawaii and drove the natives crazy by insisting that they cover themselves with cloth came from Boston and environs, where anybody who didn't cover himself with cloth caught pneumonia and died. To these cold country people it was obvious that God intended folks to cover up.

The basic reasons for the cover-up may have become confused with morality over many generations of teaching children how to take care of themselves.

"My child, you must dress properly."

"Why, Mother?"

"All decent people dress properly."

"But why?"

"Because that's the way God made things, that's why!"

124

In warmer climates, this question just doesn't come up. God expresses comparatively little interest in clothing where its absence fails to cause the premature death of a significant percentage of the population.

As Mark Twain said: "I have been asked why I dress in white. I answer that I have observed that naked people have little or no influence in Society." Spoken by a man who lived in Connecticut. The humor of this is almost certain to escape naked people who do have influence in their societies.

Recall the ability of the Aryan Indians to drink milk. That's a high-latitude trait that has become a moral issue in the southern area to which they have moved.

We have noted that the Hindus have a hard time persuading their southern neighbors to accept the northern morals, succeeding to only a small degree.

It is a general observation that no religion becomes dominant in an area where the average temperature is lower than the average temperature in the area where the religion was founded—with a few minor exceptions.

Figure 6.1 shows the data on this matter plotted from conventional sources—U.N. figures, encyclopedias, almanacs.

Obviously, we are taking a rather broad definition of religion here and making some distinctions that are not common among people who have profound convictions regarding the absolute validity of one or another of these religions. (We accept for our study whatever definition is implicit in the religious census quoted by the U.N. and the Catholic, Jewish, or Iranian almanac.)

We might have chosen to speak of "ethics" rather than of religions. The term is less loaded, but religions do tend to be the formally organized expressions of the bodies of ethical attitudes that prevail among various people.

If we speak of "formally organized expressions of the bodies of ethics that prevail" it is easy to accept Communism as one such organized expression. That language is very cumbersome and when you get right down to it, the names of the religions are the handles we use to identify the formally organized expressions of prevailing bodies of ethical attitudes. So we use the common names here.

People of ecumenical bent will be distressed to see confessional Christian religions distinguished so firmly from the Protestant religions, but again, this is just a report of the news. Certain fundamental ethical principles are different. Confessional Christians dominate countries in which the average temperature is 50–59° F (the temperature of the region where the body of ethics was first formalized) and in warmer climates. There is but one maverick in that group which is dominant in a colder country, while the influence reaches far into warmer areas. The Protestant ethic was first established where the temperature averages ten degrees colder.

Religious Dominance of Countries by Temperature

Religions dominant in countries	Mean annual temperature of country population centroids					
	Degrees Fahrenheit					
	30–39	40–49	50–59	60–69	70–79	80+
Protestant	I	ЖТ•III	I	IIII	IIII	II
Communist	I	ЖТ•II	II	II	I	I
No dominant religion		I	I			
Confessional Christian		I	ЖТ•II	ЖТ ЖТ	ЖТ ЖТ	ЖТ ЖТ
Shinto			I •			
Moslem (Shi'i)			I •	I		I
Moslem (Sunni)				ЖТ•	ЖТ IIII	ЖТ I
Moslem (other)				II	II	ЖТ
Jewish				I •		
Animist (tribal)				III	ЖТ II	ЖТ IIII
Hindu-Buddhist				I •		
Animist (Bud. Tao. Confucian)				I	I	
Animist (Roman Catholic)					II •	
Coptic					I •	
Animist (Buddhist)					II	
Buddhist					I	I
Animist (Moslem)						II •
Buddhist Taoist Confucian						I •
Hindu						II •

ЖТ II Each cipher represents one country.

Confucianism did dominate China; now Communism does.

Lamaist Buddhism did dominate Outer Mongolia; now Communism does.

A Dot is in the column to identify the temperature at which the religion was founded.

If no dot (•) is in a row, all of the ciphers in the row represent countries dominated by a religion of local origin.

Fig. 6.1.

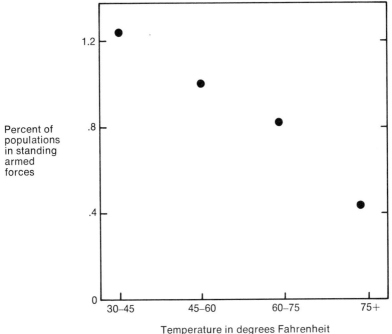

Temperature in degrees Fahrenheit
Mean annual temperatures at population
centroids (averages of 133 countries)

Fig. 6.2.

Given that average temperature is largely a function of latitude, one must ask if ethical attitude—religion—is not also largely a function of latitude. We can't answer. We can only point to the correlations.

Similarly, the attitude toward maintenance of standing armies appears to be a function of latitude. Figure 6.2 is a report of the news derived from encyclopedias, almanacs, etc. You may check these figures by looking up the information in any public library and doing your own arithmetic.

The way to get these numbers is to figure out what the mean annual temperature is at the center of population for each country. Then, for each of the countries on your list, find out what percentage of the population is represented in the standing army of that country (the "peacetime" army). Then determine the average percentage of the population in the standing army for each temperature range in which you are interested.

In this case, we had data on 133 countries. We moved in fifteen-degree steps from cold to warm.

We find that the coldest countries keep an average of 1.2+ percent of their population in standing armies. This percentage diminishes steadily as the temperature rises until, in the warmest areas of the Earth, a little less than one-half of 1 percent of the population is maintained in a standing army.

Does this suggest that things are more peaceful in the warmer regions?

Well, if you track down the rest of the data in the standard sources, you will find that the countries with the larger standing armies:

1. Start fewer wars.

2. Get dragged into fewer wars.

3. Win a greater percentage of the wars in which they become involved.

4. Lose a smaller percentage of their population to death in combat.

We may or may not agree that aggression against other nations is aberrant behavior. How about internal affairs—coups and revolutions?

There is room for infinite argument among purists about what is a coup or what is a country. We are certain that, in view of the great number of both coups and countries, a slightly different definition would not appreciably affect the results of our study.

Here again, we scoured the standard sources for reports on coups and revolutions and then plotted them according to latitude rather than starting with temperature, per se. A "country" here is defined as a political and geographical entity, recognized or de facto. This involves a certain amount of arbitrary choice, since, for example, the recent armed disagreements in Portuguese Angola are regarded by some as a war between nations and by others as a civil war. We have classed Angola as a country—a clearly defined geographical entity in which there occurred a quarrel over sovereignty by people who live there.

Revolutions and coups are defined as efforts (win or lose) to change the nature of the sovereignty. This almost inevitably involves erasure of the commitments of the old sovereignty, a wiping of the slate.

We examined 387 revolutions and coups in 153 countries all over the world in the two decades 1955–64 and 1965–74. We disposed

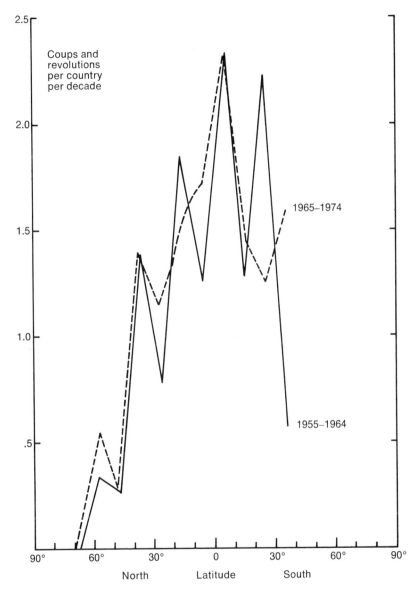

Fig. 6.3.

of weighting toward the equator by dividing the number of coups and revolutions at each latitude by the number of countries at that latitude. Thus our figures in Figure 6.3 represent the average number of revolutions and coups *per country* at the various latitudes.

The distribution does not appear to be random. If you take the frequency of coups and revolutions as a measure of stability, then stability appears to be a function of latitude. At least, the correlation is present, with or without cause and effect.

Figure 6.4 is another representation of the same information. The difference in presentation still makes the relationship to latitude very striking.

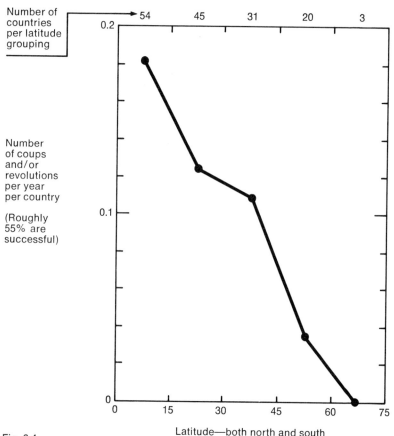

Fig. 6.4.

It may be that social stability is not ethically significant in warmer climates for the same reason that covering oneself completely with cloth is not ethically significant. Nothing outstandingly bad happens to people who fail to put a high value on clothing or large-scale stable social organization in warmer climates. That is, personal survival is not especially at stake in the short haul. There is no urgency about laying out complex social organizations that may not be needed until months or years in the future. Besides, coups are usually cheaper than elections, and may be the only change of administration that a poor country can afford.

In cold climates anybody who doesn't spend a lot of time in planning for future contingencies—filling his root cellar with potatoes, making cheese, and selecting leaders who will organize things so that the snow will always be plowed from the streets and fuel will always be readily available—is in danger of being missing from the population come spring.

Priorities vary from place to place.

We may wonder why, after spending millennia adapting ideally to a particular climate, people have insisted on moving to hostile climes and annoying their neighbors. It is obviously wiser, more conducive to survival, to stay where you know the territory and you are suited to do well, whatever you value.

Things change. We'll talk about it.

7

We spoke in an earlier chapter of the clear way in which tree rings record environmental change produced by volcanic activity. The fact is that trees are big, strong, hardy plants that have great tolerance for broad change—at least, some of them do, and those are the ones we use to obtain our information. Not all plants are as hardy as those trees. Indeed, some are notable for their sensitivity to environmental change.

Consider some rather sensitive plants:

Wheat
Rice
Corn
White potatoes
Yams
Breadfruit
Barley
Rye
Beans

Roughly 95 percent of the world's food supply is derived from this small handful of plants, along with a few others, like the squashes and common roots (beets, etc.).

If one of these plants were wiped out suddenly by a sweeping disease, the dislocation in the world's food supply would be catastrophic. (Some of us who have never seen breadfruit take the notion of its importance with a grain of salt, but it is easy to believe that the disappearance of corn would be dramatic news.)

The possibility of such a wipe-out preys on the minds of agricultural specialists, and they do what they can, not only to keep existing strains healthy, but to cultivate alternatives that might be pressed rapidly into service in case of need. The need has come up before.

After World War I an epidemic of wheat rust broke out in the United States and destroyed our soft wheat crops. We found ourselves unable to grow the traditional soft wheat upon which our milling industries were based. Fortunately, hard-wheat types were available and we began to grow a different sort of wheat. We also had to *mill* the new type, and the descriptive terms "soft" and "hard" are literal. The hard wheat is physically harder than soft wheat and we had to make changes in the machinery that handled the stuff. This is not a great strain for a highly industrial country, but real trouble for people who can't suddenly make massive changes in machinery. It cost us about $2 billion to switch the machinery.

The hard wheat has its drawbacks. It is subject to smut, a fungus that is essentially harmless, but which discolors the wheat and produces a strong odor. In processing, these unattractive features are removed and few of us are conscious of the problem as we toast our bread. The problem is noticeable to the Chinese, however, who are not accustomed to receiving smutty wheat and who have indignantly rejected U.S. shipments in recent times.

A major problem with cultivated plants is that they are carefully selected and bred for high food production in the particular areas where people use them. That is, they are tuned up to do very well under a narrow range of environmental conditions, which are controlled as closely as possible by men who irrigate, aerate, fertilize, and cultivate the plants.

Unlike the trees we use, these plants fail and die when they get too hot or too cold or too dry or too wet. No big change is good. Consistency is the key to success. Variety is the key to trouble.

Human beings depend upon these fragile plants for their lives. When the weather turns sharply variable, the plants die—and the people die. It's as simple as that.

No, take that back, it's not as simple as that. Nothing is simple. At least one more major factor in plant growth should be considered here without dwelling on it at length: the alternation between periods of light and periods of darkness.

The subject came up in fascinating context back in February 1960, when the Russian publication *Priroda* (no. 2, February 1960, pp. 124–28) discussed the work of an engineer named V. B. Cherenkov. Cherenkov had proposed a method for changing the climate in the latitude of Russia.

The Russians have been preoccupied for a long time with the notion that climate might be altered in some way to their advantage. In general, what they have in mind is warming things up so they can grow more food.

Cherenkov had a plan that called for creation of a "ring of Saturn" around the Earth. He suggested that rockets be used as delivery vehicles for dust of appropriate particle size which would be caused to orbit Earth in a great ring at fifteen hundred kilometers altitude. The ring might not only be strikingly beautiful, he felt, but would reflect wholesome amounts of life-giving sunshine onto Earth at its higher latitudes. The project was a big one. Cherenkov offered calculations that suggested that thousands of trips by rocket freighters would be required to cart the material into an appropriate orbit. He pointed out some other considerations that deserved further work, and closed with the comment: "Naturally, all this will require preliminary small-scale experimentation."

In all, Cherenkov's proposal was entertaining stuff, plausible, bold, with an air of romantic, noble purpose. A lot of people were not entertained, however, and went out of their way to stomp on poor Cherenkov.

One complaint was that nobody could see a way to maintain a "ring of Saturn" around the Earth. It was agreed that the stuff would drift out of its orbit and shade the Earth generally, bringing on effects that the critics expected to be bad.

The *Priroda* article quoted a biologist, A. F. Kleshnin, who pointed out that changes in temperature might not be the most interesting effect of such a ring. He worried aloud about the effect of nocturnal illumination. He said:

> The point is that the great mass of types of vegetation (including cultivated vegetation on which the existence of man depends) have in the course of evolution adapted themselves to a definite change of day and night. The optimal growth and accordingly the optimal vegetation yield is observed under conditions of 16–18 hours of daylight. Perpetual illumination usually lowers the yield by 20–30 percent and in some cases causes the death of plants (tomatoes).

He pointed to rice, maize, soya, and millet grass as examples of crops that will not yield unless days with not more than twelve hours of sunlight occur at some stage in their development. It seemed to Kleshnin that people who depend heavily on those crops in regions like China, India, Japan, and Africa would face serious food shortages if their days were suddenly extended by light reflected from a high cloud at night.

Even long-day crops like wheat, barley, rye, and oats require short days at certain stages of their growth if they are to flourish.

Trees, he said, are sensitive to changes in the length of the day, and not only most of our fruit trees, but species like birch and oak might well be wiped out by perpetual illumination. Kleshnin seemed to think this would be a bad thing and his remarks concluded with an extraordinarily revealing sentence: "It is possible *and necessary* [italics ours] to alter the climate, but it must be done after weighing up all the circumstances."

Fifteen hundred kilometers is a lot more than the fifty miles at which noctilucent clouds are observed, and we do not suppose that the dust ejected by volcanoes will reach such an altitude. Cherenkov's proposal does not necessarily amount to the same thing as the natural phenomena that we think we see at work, but it is clear that very high-altitude dust may alter the length of the day significantly, flooding Earth with sunlight for an hour or two before true dawn and after sunset.

That extra light may be enough to keep the plants awake, to affect the food supply significantly, and to cause man's dynasties to rise and fall. Obviously, we have not done the research and calculation necessary to evaluate this idea adequately and the matter deserves to be high on the priority list of anybody who plans to do a comprehensive study of this whole question. Perhaps A. F. Kleshnin and other researchers at the Artificial Climate Laboratory of the Institute of Plant Physiology in he U.S.S.R. have considered the matter further since 1960 and can add to our knowledge.

One plant in which people have shown great interest from ancient times is the grape. Good records have been kept of the performance of the vines and wines in Europe for hundreds of years. While wine does not supply the greater part of the necessary nutrition for the people, even in France, the performance of the vineyards reflects the general condition of the crops. A cool spring-summer makes for late harvest of the grapes and for poor quality in the wine.

The year 1600 was an especially cold time. It was the first of the maxima of the "Little Ice Age," a time when glaciers advanced to their greatest extremes in centuries. We know that this was a difficult time. It was the start in Russia of the "time of troubles." You may imagine that when some period in chronically troubled Russia sticks out in the memory of the people as *the* time of troubles, things must have been in discouraging shape. Half a million people died of starvation in Moscow alone in 1600. China, too, was ridden by famine. Northern Europe was very hard pressed.

The wine harvest in France took this pattern:

Table 7.1.

The wine harvest in France took this pattern:

Year	Date of Harvest in France in Days' Deviation from the Mean			Quality of Wine from Rhenish Vineyards −6 Max Good +6 Max Bad
	S	N	Nation	
1599	−15	−19	−17	−6
1600	−1	+8	−3.5	+6
1601	+2	+9	+5.5	+6
1602	−8	−11	−9.5	+6
1603	−14	−22	−18	−6
1604	−11	−11	−11	+2

The period was extremely variable, with wide swings in the time of harvest and with three absolutely terrible, two excellent, and one poor year out of six.

Another period may be considered with this record in mind. Seventeen-eighty-three was the year in which Franklin noted the extensive dry fogs in Europe.

Lamb states that for two months following eruptions of Eldyjar in Iceland and Asamu in Japan in 1783, the Sun remained invisible in the south of France until it rose 17° above the horizon.

For the five years following 1783, the average growth of tree rings increased south of 40° north latitude, and decreased north of 40° north latitude.

In 1783 the Chalisa famine swept India from the eastern edge of Benares to Lahore and Jammu.

From 1781 to 1789 in Japan, earthquakes and volcanic eruptions were followed by a drought known as the Famine of Temmei, and

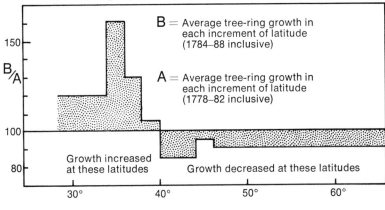

Growth of tree rings by latitude before and after volcanic eruptions of 1783

B = Average tree-ring growth in each increment of latitude (1784–88 inclusive)

A = Average tree-ring growth in each increment of latitude (1778–82 inclusive)

Growth increased at these latitudes

Growth decreased at these latitudes

Degrees north latitude

Fig. 7.1.

after this came equally destructive floods. Over a million deaths resulted.

In August 1784, a hard freeze in Mexico killed the crops and 300,000 people starved to death.

Consider the wine harvest record in France at that time.

Table 7.2.				
Year	N	S	Nation	
1778	−4	−3	−3.5	
1779	1	−5	−2	
1780	0	−6	−3	Average −2.7
1781	−3	−15	−9	
1782	4	4	4	
1783	1	−8	−3.5	
1784	−4	−9	−6.5	
1785	4	3	3.5	
1786	1	2	1.5	Average −.4 (2.3 days later)
1787	9	7	8	
1788	−5	−2	−8.5	
1789	6	7	6.5	French Revolution

The years following 1783 in France were characterized by late wine harvests from cool springs and summers, bad quality of wine, dearth of food, food riots, general unrest, and finally an extremely

137

violent revolution, during which everybody who lived in a white house was marked for execution.

And another chronology:

1660: an outburst of four volcanoes that pumped a significant amount of dust into the atmosphere.

1660–61: a large regional famine in France.

1661: one of the worst droughts ever recorded in Mexico, with heavy loss of cattle.

1661: famine started in India, with not a drop of rain reported in two years.

1662: another famine in France.

1663: violent outbreak of smallpox. Dreadful heat by day and frost at night killed crops in Mexico.

1663: Ottoman military machine conquered Transylvania, selling 80,000 captives as slaves.

1664: another volcanic duster erupted. New York was taken from the Dutch by the English.

1664–65: Newton formulated calculus, law of gravity, composition of light.

1665: Colbert, controller of France, forbade the export of wheat from France.

1665–68: the Spanish officials in Gran Quivira in New Mexico recorded three splendid years of harvests and progress at their mission. Gran Quivira was an old city of stone pueblos with a population of several thousand people. The church was impressive and beautiful. The pueblo was handsome and complex, with many ceremonial kivas.

1666: Great Fire of London.

1666: French Academy of Sciences founded.

1667–71: Stenka Razin led the whole lower Volga in a Cossack rebellion against the rule of Moscow. (Captured and executed.)

1668: a famine began in New Mexico which eventually killed more than half of the upland Pueblo tribes and forced them by 1671 to abandon the cities in which they had lived for 350 years.

The Spanish at Gran Quivira recorded local events in their notebooks. Presumably, by 1670 or so they would willingly have left the area for some better place, but it's a long way from Gran Quivira to anyplace else in a modern automobile; it's farther still on a starving horse.

The Spanish actually didn't have much to complain of and they recorded with some satisfaction that many of their belongings were made of leather which they could eat. After a couple of years of this, they did complain—not about eating leather, but about the fact that they were running out of leather.

In 1671 Gran Quivira was permanently abandoned, along with the mission pueblos Abo and Quarai in the same general region. These places may be visited now. Nobody lives there and it is hard to picture them teeming with people.

New Mexico didn't recover from this disastrous time rapidly. In 1675 it occurred strongly to the Indians that this extended outburst of trouble was the fault of the Spanish. The Indians had made it through hundreds of years without having to abandon their homes, and though the Spanish priests talked convincingly, they obviously had little influence in heaven. Logically the Indians planned an uprising, picking the very day five years ahead of time. In 1680 most of the pueblos rose and killed all the Spanish they could get their hands on—some thousands of them. At Acoma, the oldest continuously inhabited city in North America (still, today), the Indians just dropped the priests over the edge of the mesa.

Up in Santa Fe, the Indians seized the Palace of the Governors and moved in. Indeed, they turned the place into a pueblo, raising extra stories on the building.

In 1693 the Spanish sweet-talked their way back into control of the area, booting the pesky Indians out of the palace and putting things back the way they had been before 1680, more or less. This involved a massacre of the Indians, of whom some number fled— people of many different pueblos. Banding together, they formed the now Laguna Pueblo, essentially becoming a new people, a new tribe. Those 1660–64 volcanoes set the wheels in motion for all this.

The Palace of the Governors today may be seen as it was under the Spanish. Inside is an excavation, started in 1968, during renovation of the building, which revealed Indian burials, skeletons, and artifacts under the floor. These are from the time of the pueblo occupation of the building.

The descendants of the refugees have established themselves at the

Laguna Pueblo in handsome style and have learned the art of dealing with enemy governments. Their present dealings with Washington are admired for their craft, guile, and effectiveness. Uncle Sam is paying for lessons the Spanish taught in times gone by.

We note from ice-core and tree-ring records that the time from about 5000 B.C. to about 2000 B.C. was a very good time, fairly warm and fairly even. The period is known as the "Great Climatic Optimum." It occurred after the last of the great glacial ice sheets finally melted away around 6000 B.C. In the middle of that sixth millennium B.C. there was a trifling disturbance and a temporary advance of the glaciers, apparently occasioned by a great volcanic eruption in North America. However, the disturbance passed and matters quieted down for about three thousand years.

During this time the Sahara was a hospitable place, not exactly a rain forest, but good country for living things, for animals, for crops, for people.

In what is now southern Algeria there is a high Saharan plateau known as Tassili-n-Ajjer, where man has often traveled in the past and has left records of his visits. There are paintings on canyon walls in the Tassili area which give us a series of views of man from about ten thousand years ago to almost the time of Jesus. These pictures show us animals in the Sahara, giraffe, ostrich, antelope. They show us ladies and gentlemen in elaborate clothing. They show scenes of combat, of village life, of ceremonies and hunts, things that have not been seen in that region for about four thousand years.

In the earliest times, from about 8000 B.C. to 5000 B.C., the area was really quite lush, with steady rains. It dried to a grassy savannah in the Great Climatic Optimum, but supported life very well. During this time, the ancient Egyptian empires were formed (the first dynasty, controlling the regions of both the upper and lower Nile, was established around 3200 B.C.). Traditional ways of life developed which must have seemed like eternal truths to people of the time.

In about 2000 B.C. something happened. Until that time there were no great deserts in the world. We have said that the Sahara was green, and so were the Gobi, the Kalahari, the Australian desert, the southwestern desert of North America, and the Atacama and Patagonian deserts in South America.

Rather rapidly, the grasses vanished, the land dried, and everything changed. Abraham wasn't wandering· with his people in the desert just for the fun of it. His forebears lived there before it was a desert and they were undoubtedly sorry to see the green things go. They were forced to become wanderers.

At the same time, as the deserts formed, the Caspian Sea, which had almost dried up around 6000 B.C., began to fill again. The remains of ancient settlements can be seen through the shallow waters of the Caspian now. The Caspian has changed a great deal with the passage of time. It is a very sensitive indicator of variations in average rainfall, since it is a great shallow basin and rather small changes in its depth are accompanied by huge changes in its surface area.

In *Climate Through the Ages*,[1] Brooks presents a fascinating description of the evidence for very great fluctuations in the level of the Caspian through the centuries. In the fifth century A.D., a great wall was constructed to discourage the Huns from riding through the area and attacking the locals. Brooks comments that the wall extended some eighteen miles into the water at that time, while one supposes that the wall went merely to the edge of the sea when it was built.

The Caspian has dropped a long way since 1949. It has been noted that a Russian who built a vacation *dacha* on the edge of the Caspian in the thirties would in the sixties have been required to walk as much as sixty miles to get to water. That's a wide beach, even in a country where things are done on a grand scale. It may be possible by now for the Huns to ride clear around the end of that defensive wall that stuck far out into the water according to Brooks's latest report.

That's a lot of change. When the water moves miles away from a shoreline city or rises to engulf it, its inhabitants are induced to move. The changes in the Caspian are not unique. The same effects occur elsewhere, as in the Great Salt Lake in Utah.

Notice an important matter here. When the Sahara dried out, the Caspian filled up. That is, as things got wetter above 45° north, they got drier around 20° north. This reciprocal change is exactly what we saw in the precipitation/evaporation charts we discussed in Chapter Five. When things get dry between 20° and 40°, they get wet above 40° and below 20°.

---

1. C. E. P. Brooks, *Climate Through the Ages* (2nd ed., rev.). New York: Dover Publications, 1970.

Brooks shows that this same reciprocal pattern holds between the western United States and Yucatan. That is, when the pueblos suffer from drought, the Yucatan people suffer from flood, and vice versa.

Table 7.3.	
Western United States	Yucatan
Wet 500–250 B.C.; 100 B.C.–A.D. 200	Dry 500 B.C.–A.D. 250
Dry A.D. 300–800	Wet A.D. 400–850
Wet A.D. 900–1100	Dry A.D. 850–1050
Dry A.D. 1100–1300	Wet A.D. 1050–1250
Wet A.D. 1300–1400	Dry A.D. 1250–1400
Dry A.D. 1450–1550	Wet A.D. 1400–

Schoenwetter (as good a name as one may have who is involved in studies of climate) and his associates gave us more information that fits this same pattern when they studied the Chuska Valley in northern New Mexico.[2]

Schoenwetter showed that over the centuries the altitude at which trees grow in the Chuska Valley has varied by hundreds of feet. On gentle slopes and plains, this can mean miles of difference. The area has very dry grassland punctuated by cactus. At higher altitudes, piñon and juniper get closer and closer together as you rise, then blend into evergreen and aspen forest on the mountainsides.

We refer so often to New Mexico for a couple of reasons. It's partly because we live in Albuquerque and a lot of local material comes to hand, and partly because the area is great for demonstrating the points we want to make about climate and human affairs. The number of people here has always been (and still is) rather modest. It is relatively easy to trace the movements and activities of people over thousands of years, and the history is not so complex that you get lost in it at once.

If we step out in front of our office here we can look north to the Jemez (the Valle Grande volcanic crater we have already mentioned) and south to Ladrones Peak. We can see very much beyond these landmarks to other mountains, but these two have special meaning

2. A. H. Harris, R. Schoenwetter, and A. H. Warren, *An Archeological Survey of the Chuska Valley and the Chaco Plateau, New Mexico, Part 1.* Albuquerque: Museum of New Mexico Press, 1967.

for us, since they are just about a hundred miles apart and we can see in one glance the whole terrain between them.

A hundred miles is just about the distance between Athens and Sparta. That is, the terrain we can see is as large as the whole area in which the Peloponnesian wars were fought with such fanfare. It is remarkable that such attention has been focused for so long on quarrels over a piece of territory that may be viewed in a casual glance.

Back to the trees—driving home from the office on the volcanic West Mesa, one must cross the Rio Grande valley and river. (In 1841 the Rio Grande was reportedly a mile wide at Albuquerque. Today, partly because of water loss to irrigation, the river isn't much more than a large brook.) The valley is full of cottonwoods and poplars and other thirsty things. As one drives up 250 feet of altitude from the valley to the house, the trees all vanish and dry grass takes over, with cactus, tumbleweed, and other sharp things. (Everything that lives here is either spiky or venomous or both, and only a newcomer tries to weed his garden without gloves.)

At the top of the mesa, above the layer of lava that flowed out here a few thousand years ago, the grass is spotted with occasional piñon. The whole increase in altitude must not be more than three hundred feet from the river to the top of the mesa, but the difference in vegetation is very distinct.

When Schoenwetter tells us that the trees moved up and down the slopes in the Chuska Valley, this is the sort of difference he's talking about. We have pointed out the periods he discusses on a plot of the temperature indicated by a Greenland ice core. It is clear that when the temperature dropped in Greenland, the Chuska Valley got a lot more moisture. This would surely bring the trees down the slopes.

More than that, it would make it possible to grow food on the mesa. That isn't possible now, even, as some of us are well aware, with the investment of a lot of energy and water.

We know a Lebanese gentleman who came to the United States in the early part of this century and became a trader with the Indians over near Mount Taylor, forty miles west of Albuquerque across the mesa and the Rio Puerco.

He comments that people used to raise corn out there by dry farming without irrigation. Intelligent young people look at the area, shake their heads, wink, and make smart-aleck comments about how the old man is losing his grip and misremembering things. Nobody

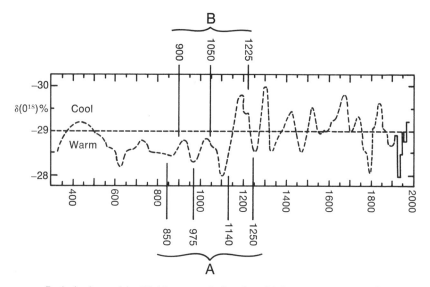

Periods A are identifiable as periods when high-energy summer thunderstorms dominated the weather pattern. Comparison with Greenland ice temperature records ($\delta(O^{18})\%$) shows these times to have been warm. Such rainfall is minimally effective and maximally erosive.

Periods B were cool or cold, with more rainfall, dominated by winter drizzle, maximally effective, causing aggradation instead of erosive degradation.

After Ladurie 1971; after Dansgaard et al 1971, pp. 37-56
Dates A and B from Schoenwetter 1967

Fig. 7.2.

could *possibly* have grown corn out there a mere sixty or seventy years ago.

Still, times have changed very dramatically in this century. Look at the temperature plot of the last eight hundred years, derived again from the Greenland ice.

It may be seen that the period of the 1930s to the 1950s was by far the warmest time in those hundreds of years, and in fact was the warmest period in thousands of years. We know from correlations of this plot with Schoenwetter that the climate in northern New Mexico is accurately represented by the Greenland core.

This would show that there was a much cooler, wetter time in the early part of this century, which our friend would remember. If he says they grew corn, we're prepared to believe it.

Look back still further. The Spanish got here about the middle of

144

the sixteenth century, when the climate was moderate. Later in that century, things got cooler. By the middle of the seventeenth century, things changed even more sharply. The weather was extremely variable in this period and the 1660s finally did in the pueblos like Gran Quivira which were not located in the Rio Grande Valley or in other favored places.

The changes in the mid-twentieth century were equally dramatic, but with the average temperature high instead of low.

The effect of this is visible in various ways. For example, the town of Estancia over to the east of the Manzano Mountains was the site of a great rancho that flourished for something like three hundred years as an agricultural center. Estancia has been especially noted as a supply point for the pinto beans that are a staple of diet in this area. In the 1970s Estancia is just like many another small city whose young people are leaving it. The bean warehouses are closed and the place is more like a relic than a beehive of activity.

All agriculture has been irrigated in the area for many years. It

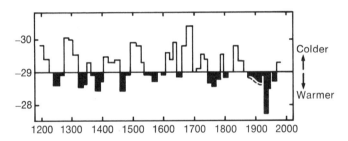

The warmest period for a long while has just passed. The Earth's temperature is returning to "normal."

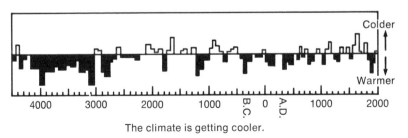

The climate is getting cooler.

After Dansgaard et al, 1971

Fig. 7.3.

wasn't irrigated in the days of its glory as a great rancho. Something has changed dramatically and the effects are obvious to the naked eye.

Southern New Mexico was real Old West country. Billy the Kid, Pat Garrett, A. B. Falls, Elfego Baca, Bat Masterson, and a great many more of the famous Old West characters flourished there in the late 1800s. The famous Chisum cattle ranch was based there. The area was famous largely because its inhabitants were extraordinarily touchy and tended to settle minor disagreements with gunfire. Dedicated buffs of the Old West may know more about this than we do, but we've never seen any explanation for this wild time other than the obvious explanation that it was the frontier and civilization had not yet blessed or cursed it.

A couple of hours in the Aubuquerque Public Library gave us a better reason. We dug out the rainfall records of El Paso, Doña Ana County, Lincoln County, Las Cruces, and Elephant Butte and took a look. The fact is that the average rainfall diminished in that area by *30 percent* between 1880 and 1930. No wonder everybody was tense and quarrelsome. No wonder the cattle business faded away. No wonder Pat Garrett became a promoter for an irrigation project in the Pecos Valley before his fellow citizens murdered him in 1908. Things were getting tougher and tougher in the area all during the famous period.

Similarly, one may visit the Pecos National Monument east of Santa Fe (a couple of hundred miles north of the area that Garrett was hoping to irrigate) to see the remains of a pueblo and the inevitable mission church.

This pueblo, too, was abandoned during a time of drought and stress. The ruins of the church were a landmark to travelers on the Santa Fe trail during the early part of the last century. Today those massive adobe ruins are still very prominent.

The pueblo was a big one, built on a ridge along the Pecos. In its heyday, the stone and adobe buildings were five stories high with hundreds and hundreds of rooms, many kivas, and excellent access to both water and farmland. According to the literature handed out at the monument, things grew very nasty there when drought limited the food supply and drove large numbers of Apache and Cheyenne into the area in search of anything that was loose and edible.

The pueblo was a fortress, defensible enough, but things got so bad that its inhabitants found it difficult to get down to the river and back with water. Farming was limited to the area *inside* the pueblo and right next to it on top of the ridge. Looking up at the top of the nearby mesa, the people in the pueblo could see their enemies watching them. Looking down at the pueblo from the mesa, the attackers could see exactly how things were going in the pueblo.

At last, the defenders crumbled and fled.

Two things about this story startle the visitor. For one thing, the river is literally a stone's throw from the pueblo. It isn't any half-mile hike to the riverbed. The ridge runs right down to it. If the defenders of the pueblo couldn't travel that distance and back without combat, they were under very close siege indeed.

The second thing is that the Pecos in that area is completely dry now, at least within a few hours after any rainstorm. Looking at the whole thing, one can hardly imagine that an appreciable population lived and farmed there for hundreds of years.

Corn on the west mesa. Corn at Pecos? It must have been so.

The point here is that New Mexico is *marginal* crop-growing country. Even comparatively small changes produce large effects, good or bad. If the average rainfall is a modest twelve inches annually and that drops to eight inches, as it did in Billy the Kid country, the effect is very impressive. It makes the difference between life and death. Those changes bring wanderers to the doors of people who have not yet been forced to move. The wanderers want food. Given a choice between fighting or watching their children starve, people lean toward fighting. The trouble cascades from the marginal areas, where the trouble rises first, to the more productive areas where people have what the wanderers need. Tempers fray.

In general, it appears that in long periods of good times, with variability at a minimum, people and societies are relatively stable. Even the nomads tend to join the stable populations. Cities and empires plod along comfortably in these times, growing ever more organized.

In *Peru Before the Incas*, Edward P. Lanning points out that all civilizations share a set of special characteristics:[3]

1. Subsistence based on intensive agriculture, with or without animal husbandry.

3. Edward P. Lanning, *Peru Before the Incas*. Englewood Cliffs, N.J.: Prentice-Hall, 1967.

2. Relatively large, dense populations.
3. Efficient systems for the distribution of foodstuffs, raw materials, and luxury goods over fairly large areas.
4. A diversity of settlement types, including either cities or ceremonial centers as the focus of sociopolitical organization.
5. State structures, with central governments exerting varying degrees of control over the lives of people in many settlements.
6. Intensive social stratification.
7. Extensive occupational specialization in which only a part of the population is involved in the production of food, while the remainder are craftsmen, administrators, merchants, and so forth.

By these descriptors, civilization calls for increasing organization, more elaboration of mechanisms that increase social stability. Civilizations constantly increase their *dependence* upon stability. The centers of population, the cities, depend wholly upon an uninterrupted flow of food and material from the countryside. If the flow is restricted for any reason, the people in the cities have a problem that cannot be solved in an orderly fashion. There is no way to grow food for ten million people inside the boundaries of New York City. No amount of administrative effort or committee meetings, of declarations in favor of free speech and natural-right-to-have-food will cause food to grow in appreciable quantity in New York. Two approaches are open. New Yorkers can go out and get food somewhere or they can die off in sufficient numbers so that the demand is reduced to fit the supply.

In good times, climatically, the civilized people dominate the world. With their numbers and with their systematic ways they can expand into marginal lands and take control of everything there. Not that there *is* much there.

The people who normally live in marginal areas like New Mexico and Mongolia and the outback and the Andean Altiplano don't mind going along with the civilized people, after some initial discussion about sovereignty, which is usually settled with a small war and a face-saving agreement not to continue the unpleasantries. As time passes, the people on the fringes of the civilizations forget about their ancestors' urges to storm redoubts and harass the outposts from the rear. Life is much more simple and pleasant for them if they just send tribute to headquarters and play the game.

The civilized folk, meanwhile, are so busy trying not to upset the system that they forget what it is like to deal with surprises. They eliminate surprises by legislation and it works out very well when the climate is peaceful.

When the climate changes, the first people affected are in the marginal areas, where the crops fail almost at once. The civilized system is prepared for this first-round trouble, and supplies are doled out from more favored areas where there is enough to share.

If the change persists, the problem begins to spread toward the center of the civilization where all of the good cropland is. The hard-pressed folk around the edges begin to resent demands for tribute. They complain that the supplies of food are inadequate to maintain them in the style to which they have become accustomed. They find that the characteristics of civilization—those we have identified above —no longer apply to their lives. There is no survival value in those characteristics. The buffer states around the empires grow restive. Revolts break out. The peripheral people polish up their old family weapons and begin to raid the better-off areas. As time passes, the people in the marginal areas have to move.

They become nomads once again, migrating as they must, seeking what they need to live. The increased harshness of their lives gives them an uncivilized approach to the world. They seek not stability, but action that will give them more of what they need. They try new ideas, finding no solace in tradition.

Bit by bit, the rough, tough nomadic people, the barbarians who won't play by the rules, encroach upon the civilizations and eventually seize them. The civilized people are always more numerous than the nomads, but they always lose during the bad times.

A few mean, cantankerous people who can not only deal with surprises, but create surprises for the civilized people, always seem to have an advantage.

Suppose that New York City, that center of good things, were to enter a serious conflict with upstate New York. New York has the population to beat the country people, but chances are that most New Yorkers have no idea where their food comes from, how much backlog is stored, where their fuel comes from, where their water comes from. If a handful of folk in the Mohawk Valley thoughtfully turned off New York City's power, blocked the railroads, scattered roofing nails on the highways, and waited a few days, they could then defeat New York City with spears and arrows.

Lanning points out that a society must have all seven of the identified characteristics to qualify as civilized. We may take comfort in the fact that the United States is *not* civilized and has not yet forgotten how to deal with surprises. Consider the points:

1. Our agriculture is not intensive, but *ex*tensive. In really civilized countries like India, China, Japan, Egypt, food is grown in every possible cranny. The Japanese grow crops in the railroad rights-of-way, sometimes between the ties, jumping out of the way of passing trains. All of the land is put to use.

   In the United States, we still have not only the wide-open spaces but little open spaces. We haven't begun to feel the pinch of life-or-death agricultural cultivation. Those who remember World War II victory gardens have a mild impression of the resources that are still available to us if we really need to make the most of what we have, but the United States has never seen anything like intensive agriculture of the sort upon which the truly civilized countries have been dependent.

2. We don't have relatively large, dense populations, compared with the rest of the world, and especially with the civilized countries.

3. Yes, we do have efficient distribution systems. We must admit the danger of civilization in this area.

4. Our cities and ceremonial centers are *not* at the focus of our social-political organization. We focus on *people*. When the President moves, his powers go with him. If he speaks from a mountaintop in Maryland, we take him as seriously as we would if he were in the White House. If he speaks from a Western ranch, he's still the President.

   If Washington, New York, Chicago, and Los Angeles were to vanish some night, most of us wouldn't object. Nobody waits for orders or for permission from those places to go ahead and do whatever seems appropriate. Indeed, a large part of the population expresses daily the wish that Washington would blow away. There would be some paperwork to clean up if it did, but the absence of Washington or any of the other great cities in this country would affect the general activities of our citizens comparatively little. In civilized countries, control of the capital is more important.

5. Yes, we have state structures that affect our lives.

6. We do not have intensive social stratification. One minority group has thought that intellectual credentials, degrees, certifi-

cates, and extensive education place them in a favored position to guide the destinies of lesser folk. They struggled in the colleges during the last decade for control of policy and the physical plant. The intellectuals came close enough to controlling the educational systems so that the people who pay for education became alarmed. While during the fifties school bonds passed in local elections, regardless of merit, they have been failing to pass, regardless of merit, since the intellectual outburst in the colleges. State legislatures are questioning all educational expenditures. Professional intellectuals, who make a living almost exclusively on tax money provided to them through schools, found that as soon as they were identified as a power-seeking class, their victuals were diminished.

7. Yes, we have extensive occupational specialization, *but* occupational mobility is very great. Even farmers are not necessarily born into their trade in our society. A tiny fraction of our population is actively engaged in growing food and fiber, about 2.8 percent. People not only leave the farms, but choose to go to them from other occupations. Occupations are not hereditary.

If Americans look around to find out where the barbarians are, wondering where the unpredictable people will come from, they need only look in the mirror.

Famine is a major driving force in human migration and social change. It is logical to suppose that climatic change that produces famine may be detected by examination of the records of famines through history. We have made a stab at this and the results are not very satisfactory. The records of famines are somewhat uneven.

Figures 7.4 and 7.5 are fragmentary pieces of work that indicate real correlation between famines and tree-ring growth over a period of 1,400 years.

The uppermost line of Figure 7.4 indicates the percentage variability in the growth of tree rings in the California sequoia series. Plotted on that same graph are the reported non-Chinese famines by century in the same period. We are edgy about the famine figures and assume that the earlier reports cover but a small portion of the famines that actually occurred. It is fairly safe to assume that the reports diminish in quality with age. Even so, the occurrence of famines appears to correlate very closely with variability of tree-ring growth.

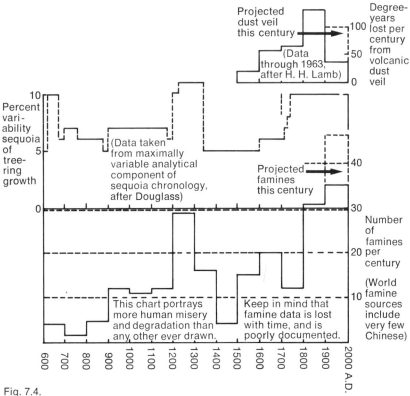

Fig. 7.4.

The period since A.D. 1500 is fairly well covered by reports of volcanoes, and it is in 1500 that Lamb's admittedly incomplete record picks up. The Lamb data were used to produce the trace that appears in addition to famine and tree-ring data. The volcano curve represents degree years lost per century from volcanoes—a measure of the lowering of temperature by dust veil.

Figure 7.5 is plotted on the same chronological scale and represents Chinese famines.

These graphs have not been worked over and massaged to improve the similarity of their form. Even so, there are correspondences in the data which lead the thoughtful observer to suspect that famines, tree-ring growth, and volcanic activity are significantly related.

One of the problems, obviously, is that a really effective famine leaves nobody to report on it. Aid officials in Ethiopia recently have reported coming upon villages in which the total population had died.

Among nomads who leave little in the way of buildings for archeologists to examine, record-keeping for posterity has chiefly taken the form of oral traditions, sagas and the like, which professional or amateur rememberers can recite on command. When famine or plague takes the rememberers, a gap is created in the historical records which cannot easily be filled.

Further, hunger takes most people's minds away from concerns like record-keeping. People who are personally deeply involved with famine are not given to formation of historical societies that will keep all the facts straight so that they can be discussed later. We have listed hundreds of famines, but we have no confidence in the comprehensiveness of the data. The records don't make light reading.

We've mentioned a famine in which a half-million people died in Moscow alone. By the standards of India, that's a trifle. During World War II a famine occurred in Bengal which might have been alleviated to a large degree if anybody outside the district had known about it and had thought to send help. The reports say that com-

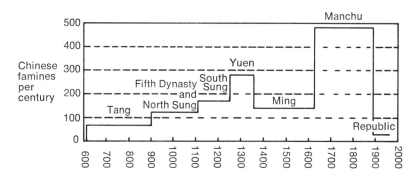

Some Particulars: From 1200 to 1368 (the conquest and rule of the Mongols [Yuen]), the population is estimated to have decreased from 100 to 65 million.

In the few years from 1800 to 1815, the population is estimated to have decreased from 430 to 380 million. We know that China's neighbor, Korea, lost 30%—2,250,000—in their famine in 1812–13.

A U.S.S.R. broadcast on April 7, 1969, said China's Communist government executed 26,300,000 from 1949 to May 1965.

Fig. 7.5.

munications with Calcutta were poor and nobody outside realized how bad things were until several weeks had passed after the problem broke out. This little oversight cost a million and a half lives.

Bengal has been a difficult place for a long time. In 1769–70 a famine occurred which killed ten million people, fully a third of the total population in the area.

Westerners are not moved as they might be by news of famine in distant places. We are so accustomed to hearing of these events in the mysterious East that they lose their meaning.

Come closer to home. Between 1846 and 1851 the "potato famine" raged, especially in Ireland, but it also affected most of northern Europe. An estimated one million people died. Interestingly, the *Encyclopaedia Britannica* provides an estimate of the size of the potato famine, not in terms of the number of lives involved, but in terms of the amount of money provided by the British government for relief—£10 million. This bureaucratic viewpoint is characteristic of the British reports on India, too. A truly civilized society can usually assure the survival of administrative functionaries in its colonies whose food supply is guaranteed and who may take a comparatively detached view of the proceedings. In China over the millennia it has been customary for famine relief by the central government to take the form of forgiving the taxes of the affected provinces. It is curious to note that though the capital may move about with time, it is the provinces closest to the capital that most often are forgiven their taxes. One might conclude from this that the crops are blighted by proximity to the seat of government, but the suspicion is that bureaucrats are more responsive to trouble that is very close to them than they are to problems that are merely reported from remote provinces.

In 1916–17 700,000 people in Germany died of starvation. There was a war going on at the time and a gratuitous potato blight was part of the problem. It is easy to shrug off these wartime disasters as man-made problems, but we are not certain that the war and the famine are not both effects of a more fundamental cause.

We observe that the terrible mud in Flanders which is so well remembered by the veterans of World War I was a gift from the volcano Katmai in Alaska, which fired its dust into the upper atmosphere just in time to precipitate all that rain in Europe.

Repeatedly in Europe there have been reports of cannibalism and of people found dead with grass in their mouths.

In Charlemagne's time, the peasants were driven by famine to eat members of their own families.

Between A.D. 1005 and 1016 half the population of England died.

Between A.D. 1064 and 1072 there was a drought in Egypt, accompanied by a failure in the flood of the Nile, upon which agriculture there depends (leading one to wonder about the effect of the Aswan High Dam in modern Egypt). Not only was there widespread cannibalism, but an open market for human meat was in operation. The butchers lowered great hooks from windows to snag passers-by from the streets and drag them screaming into their lairs. The victims were slaughtered, dressed, cooked, and offered for sale in shops. Descriptions of the same kind of dramatic activity color accounts of the seige of Leningrad during World War II.

Between A.D. 1315 and 1317, 10 percent of the population died in central and western Europe because of crop failures associated with excessive rain, not drought.

The years 1452 to 1454 were difficult in North America. There was general cannibalism in Mexico. Children were sold for a few handfuls of maize—to be sacrificed and, in some instances, eaten. It is estimated that fully half of the population died in this time. The problem again was not drought, but floods and great cold.

It is odd, as we look back on it, that the history of mankind should seem to us like a smooth continuity disturbed only incidentally by ripples of travail. In fact, it is surprising that we have preserved so much of the past—knowledge, attitudes, traditions, language. People are pressed to incredible extremes of hardship. Their numbers are slashed away time and again until barely a shred of a society may survive to represent the old times to the future. The course of history is not smooth, but occurs as a series of abrupt lurches and staggers with occasional periods of steadiness.

In Harold Lamb's biography of Charlemagne a striking image is that of the Frankish king on his travels through Europe, passing through the ruins of ancient cities. Paris, the "City of Light," whose residents speak with pride of its long history, was wholly abandoned. It was a ghost town, a puzzling symbol of the good old times to Charlemagne. Europe was filled with such specters, largely the remains of the Roman Empire.

The cities of Mohenjo Daro and Harappa on the Indus River are as ghostly a pair of towns as you may hope to find. At one time they were big cities, clearly the centers of substantial societies, yet they were snuffed out completely long ago, leaving us little by which we can figure out who the people were, what they were like, or where they went. We are quite certain that the Aryans who brought Hinduism and milk to India were the people who succeeded the original proprietors of these great cities, but many mysteries remain. For one thing, the cities are in a hot, dreadful wasteland that cannot support a large population. It is certain that the area has changed since the cities were constructed around 2800 B.C. Some observers have speculated that the inhabitants chopped down all the local forests and did themselves in by changing the place to a desert. That's an ambitious project. We're inclined to bet on other causes.

Another factor is especially strange about these cities. Not only did they come to an abrupt end; they had an abrupt beginning.

In cities like Babylon, Troy, and Thebes, one digs down through layer after layer of ancient city which becomes increasingly primitive with the depth of the dig. At last, one comes to virgin soil that bears no more trace of human activity. It is obvious that one has dug back through a long history of human development from sophisticated to very primitive.

At Mohenjo Daro and Harappa one digs through the sophisticated levels and finds—virgin soil. There is nothing primitive there, no long record of development and trial and error. Apparently the people who built those cities traveled to the sites from someplace where they already knew how to build cities and promptly built a couple. The whole technology was transplanted intact to the new site. That's a very impressive situation—obviously, a large number of highly skilled people made a coordinated move. Why?

People who admire civilization have traditionally assumed that everything of value is developed in the civilized centers and then permitted to filter out from the cultured folk to the largely ungrateful barbarians. We speak here of domestication of plants and animals, mechanical inventions, architectural designs.

It ain't necessarily so. In a 1971 *Scientific American* article Colin Renfrew[4] pointed out the great significance which the correction in carbon 14 datings had with respect to the notion that all good things diffused out to the hinterland from points of origin in the ancient civilizations of the Near East.

4. Colin Renfrew, "Carbon 14 and the Prehistory of Europe," *Scientific American*, vol. 225, no. 4, 1971.

Renfrew uses as examples structures like Stonehenge, megalithic passage graves in Brittany, designs in stone on Malta, fortifications in Spain, and metallurgy in the Balkans. It is apparent, since the carbon dating system correction, that these things were developed earlier in the barbarian areas and in some cases traveled *to* the civilized centers, which systematically copied and improved upon them. The Phaestos disc was printed with movable type in 1400 B.C. in Anatolia. The wheel was developed around 3500 B.C. in the Steppes. The stirrup, too, came to civilization from the Steppes. Parallel development is a possibility in some cases, but in any event, the barbarians made their own innovations.

Some influence, other than highly organized government, impels people to do work that appears in retrospect to be desirable. Something makes people restive, eager to change things, eager to move— at least *willing* to move.

We haven't come this far without making it obvious that we suspect climate of being the driver behind human affairs. Wherever we look, we find bits and pieces of evidence that this is so. Until the massive work of correlating all this material can be done (perhaps over the next twenty years), all we can do is stare at the pieces.

They make interesting viewing. Consider a few.

The chart which follows is based on variations in the average temperature in the United States from A.D. 1500 to almost the present. Against this curve we have plotted all of the major conflicts that have occurred *within* the continental United States between 1600 and the present. One may be confident that the list is incomplete, but it represents somebody's best estimate of significant matters.

The times of great temperature change, therefore times of great variability, are marked by combativeness among the residents of the United States. That great dip in things marks the time of the Civil War, during which we discussed our differences with unprecedented ferocity.

By the way, while our Civil War was killing 850,000 of us, China was getting what entertainment it could from the Taiping Rebellion (1851–64). Wrapped up in the importance of the War Between the States, we don't pay much attention to these little foreign squabbles. Few Americans are aware that the Taiping disagreement, led by a chap who believed he was the brother of Jesus Christ, killed twenty million people.

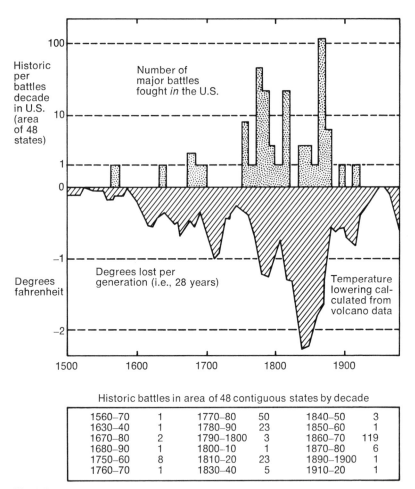

Fig. 7.6.

Historic battles in area of 48 contiguous states by decade

1560–70	1	1770–80	50	1840–50	3
1630–40	1	1780–90	23	1850–60	1
1670–80	2	1790–1800	3	1860–70	119
1680–90	1	1800–10	1	1870–80	6
1750–60	8	1810–20	23	1890–1900	1
1760–70	1	1830–40	5	1910–20	1

Actually, the lower curve in Figure 7.6 is not just a simple record of temperature, but a more complex calculation that ties temperature more closely to human affairs.

We have assumed that one human generation is twenty-eight years, on the average. In a generation, a person accumulates the information and attitudes that will be passed on to the next generation. Assuming that climate has something to do with attitude, we constructed this chart by calculating for each year the *average* temperature at which a person then twenty-eight years old would have spent his life.

Thus, in 1780, a twenty-eight-year-old person would have lived in an environment that was 1.23° F below normal. This would be equivalent to his living at a latitude an average 369 miles north of the true latitude.

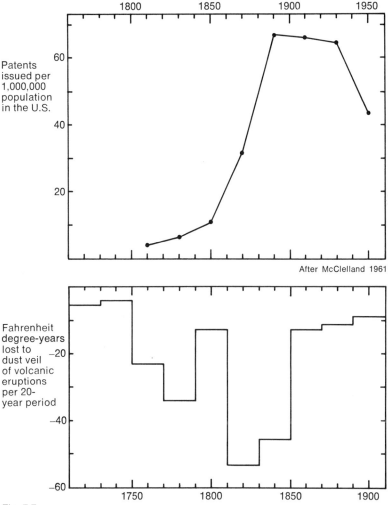

After McClelland 1961

Fig. 7.7.

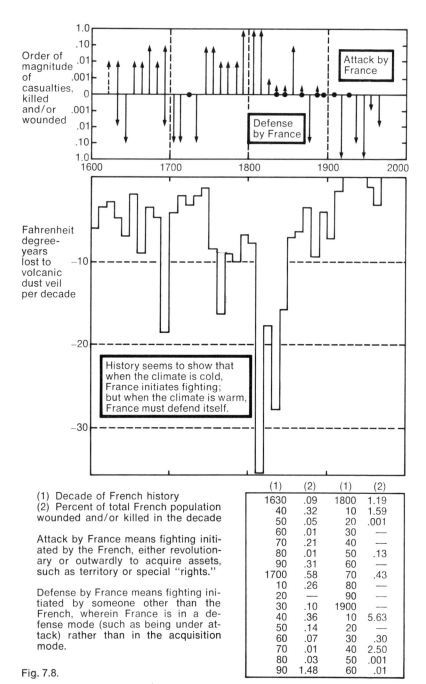

(1) Decade of French history
(2) Percent of total French population wounded and/or killed in the decade

Attack by France means fighting initiated by the French, either revolutionary or outwardly to acquire assets, such as territory or special "rights."

Defense by France means fighting initiated by someone other than the French, wherein France is in a defense mode (such as being under attack) rather than in the acquisition mode.

Fig. 7.8.

(1)	(2)	(1)	(2)
1630	.09	1800	1.19
40	.32	10	1.59
50	.05	20	.001
60	.01	30	—
70	.21	40	—
80	.01	50	.13
90	.31	60	—
1700	.58	70	.43
10	.26	80	—
20	—	90	—
30	.10	1900	—
40	.36	10	5.63
50	.14	20	—
60	.07	30	.30
70	.01	40	2.50
80	.03	50	.001
90	1.48	60	.01

A twenty-eight-year-old native of Columbia, South Carolina, in 1836 would have spent his life in a climate the same as that of Toronto, Ontario, in 1936. A person who had been living for twenty-eight years in Louisville in 1840 would have experienced the same climate as a twenty-eight-year-old in Stalingrad in 1940.

We tend to think of the Southerners who fought the Civil War as slow-speaking, deliberate people who paced themselves thoughtfully. In fact, the guys who fought the Civil War for the South were not hot-country people by our present standards. They were hard, tough cold-country types.

Times change.

There is a deal of evidence that indicates that people do in later life what they have been taught as children. That is, all imprinting of basic convictions occurs by the time a child is seven. Each of us operates on these beliefs and our view of reality is filtered through our childish perceptions.

Given this, the things we do as adults in control of society fifty years later may have nothing to do with the realities of the time, but a great deal to do with what we perceived before the age of seven.

Figure 7.7 shows the average temperature of the United States by decade from 1740 to 1920. Above that is the average rate per capita at which patent applications in the United States have been filed between 1810 and 1960.

The rise in the rate of invention clearly follows the decline in temperature—but note that the scales are offset by fifty years. Things warmed up a great deal during the 1930s. One wonders if our rate of innovation will plunge dramatically between 1980 and 2000.

For a last tidbit, look at this comparison of the war casualties per decade in France between 1600 and 1900 and the degree-years per decade excursion from nominal calculated from dust veil between 1600 and 1900.

Why do people leave the places for which they have grown best suited by nature. Why do they quarrel and break the peace? Why do they fling themselves upon their neighbors.

It appears that we are all driven by pressures over which we have very little influence. When the tidal forces persuade the volcanoes to speak, the volcanoes say: "Jump!"

Our only question is: "How high?"

8

Since we are much concerned with geography in this discussion, especially latitude—the distance of a given place from the equator—it is probably worthwhile to review some of that geography and recall where things are. It's easy to forget.

One of the problems, for Americans at least, is that almost all of the maps of the world we see are Mercator projections. All world maps try to represent the surface of a sphere as best they can on a flat piece of paper, and all representations distort the surface of that sphere. They have to; there's no escape, though one is permitted to choose the distortion he likes.

The Mercator projection is a dreadful distortion, especially at the high latitudes, but it does spread the sphere out very neatly so that everything at the same latitude is distorted equally and the latitude and longitude lines are all straight. That gives us the illusion, probably, that Mercator is easiest and best for most purposes.

The distortion at high latitudes is so very great that the representations of the polar regions are useless and the top and bottom of the projections are almost always chopped off. Since most of the land mass of the world is in the Northern Hemisphere and a useful part of the scene is above 75°, the map usually runs up to about 80° in the north.

On the other hand, there is very little of significance in the Southern Hemisphere beyond 55°, so the bottom part of the map is extended to only about 65° south to include the tip of the Palmer Peninsula in Antarctica. This is all very sensible and convenient.

However, it makes for a persistent and peculiar impression in the minds of the people who are exposed to these maps from childhood.

Look where this arrangement puts the equator, well below the middle of the map. In general, the equator is placed more than six-tenths of the distance from the top to the bottom of the map, sometimes even seven-tenths. This arrangement imprints in the mind of the casual observer, whose livelihood does not depend on his knowledge of geography, a profound impression that the area in the middle of the map is at the equator. That's where you expect the equator to be.

People who think a lot about geography may not have a problem with this distortion and disorientation, but the rest of us have to look at the map whenever we want to come within twenty degrees of the proper latitude of a named place.

Now, without peeking, try some estimates yourself. We have made a big point of saying that when the climate gets cooler, the area between the equator and 20° gets effectively drier; and zones from 20° to 40° get effectively wetter; the zones from 40° to the poles get

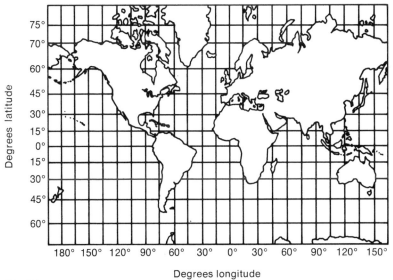

Fig. 8.1.

163

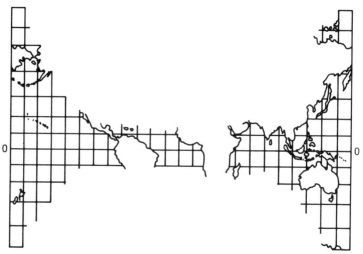

Fig. 8.2.

effectively drier in *general*. (This "effectively" business has to do with the fact that, while rainfall is increased all the way up to 45°, the lower temperature counteracts the good effect of increased moisture on plants. Since plant growth is our primary concern, we concentrate on the 20–40° zone.)

Where is the Sahara centered? It's a hot place spotted with oases full of palm trees. Equatorial? No. It's centered just about at 23° north, right on the Tropic of Cancer. It doesn't come within twelve degrees of the equator, except at its eastern end, where it sneaks down along the edge of Ethiopia, through Somaliland to Kenya, and almost reaches the equator. Most of the Sahara is in the region that gets more rain when the planet cools. However, the southern regions of the Sahara get drier.

How about China, home of bamboo and coolie hats for protection from the hot sun? China is a big country and it runs a long way, north to south, but the only part of China that extends south of 20° north latitude is the island of Hainan, over which we had some nervousness during the unpleasantries in Viet Nam. Most of what we call China lies between 20° and 40° north. The northernmost part of "China" (Manchuria) runs up to about 53° north. All in all, China is in virtually the same zone of latitude as the United States.

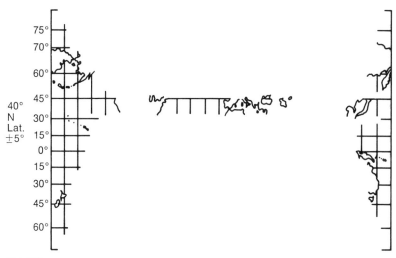

Fig. 8.3.

People in the United States think of themselves as having predominantly a European heritage. The early settlers, English, French, Spanish, named towns in the New World after towns in the Old.

London, England, is about 52½° north. New London, Connecticut, is about 40½° north. That's twelve degrees of difference, a huge amount, almost eight hundred miles. London is in the southern part of England. This places all of that country far above the 40° precipitation/evaporation/temperature crossover.

Madrid, capital of sunny Spain, is at about 40½° north, a trifle north of the center of the country, whose southernmost point is about 35° north. Madrid, New Mexico (a place of little distinction unless you like ghost coal-mining towns), is about 34½° north.

Orleans, France, is almost 48° north. New Orleans, Louisiana, is almost 18° south of that. Even Orleans, Canada, is a couple of degrees south of Orleans, France.

Tropical paradise Honolulu is about 21° north.

Buenos Aires is about 35° south.

Rio de Janeiro is about 22° south.

The country of Chile runs clear from about 18° south to almost 55° south.

Johannesburg, South Africa, is about 27° south.

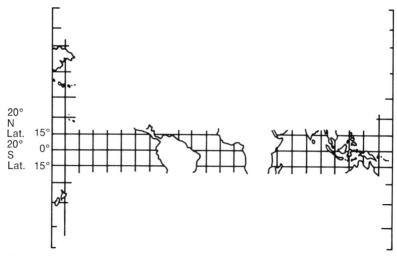

20°
N
Lat. 15°
20°   0°
S
Lat. 15°

Fig. 8.4.

The "warm" Mediterranean Sea runs from about 30° north up to almost 44° north at Trieste. That's farther north than Toronto, than Milwaukee, than Boise.

Start from the international dateline and keep heading west on the 40° north latitude line and tally the great cities that are within five degrees of that line, remembering that 40° is the crossover point at which there is very little change in growth conditions with changes in average temperature. Tokyo, Seoul, Peking, Tientsin, Tashkent, Samarkand, Teheran, Baku, Ankara, Istanbul, Athens, Bucharest, Sofia, Belgrade, Naples, Rome, Genoa, Marseilles, Algiers, Barcelona, Bordeaux, Madrid, Lisbon, Boston, New York, Philadelphia, Pittsburgh, Chicago, St. Louis, Denver, Salt Lake City, San Francisco.

Obviously, these cities are not all comparable, but they are the ones that leap to the eye if you follow the line. Clearly, many big, important cities are north of this band, many south.

Follow the equator and see what's within five degrees on either side. Um . . . Halmahera? Um . . . Let's make the band wider at the equator, since there isn't really all that much land mass at that latitude. Spread the band out to ten degrees north and ten degrees south. Back to the international dateline and start again, heading west. Um . . . Port Moresby? Halmahera? Davao? Djakarta, Singapore, Kuala Lumpur, Columbo, Addis Ababa, Nairobi, Kinshasha, Lagos, Acra,

Monrovia, Recife, almost Caracas, Bogotá, Quito, San José . . .

Even if you spread the band out an additional twenty degrees, so that it is forty degrees wide, reaching from 20° north to 20° south, your sweep will pick up only a few more (albeit important) cities whose names the average reader will recognize. Manila, Saigon, Bangkok, Madras, Bombay, Khartoum, Port Guinea, Dakar, Brasília, La Paz, Lima, Managua, Mexico.

This is not meant to be a comprehensive list, just a reminder that the population of the world is not distributed uniformly about the globe and the real distribution is a surprise to almost everybody.

Here's another that throws most Americans: What is the population of Canada, our big old Anglo northern neighbor?

And what about Mexico, our *amigo south of the border?*

In the early seventies, the population of Canada was about 20 million. The population of Mexico was about 47 million. Most Yankees estimate exactly the opposite of the real numbers. It seems hardly fair to Mexico.

Jerusalem, to pick a place at random, is almost 32° north.

Latitude is not the only factor involved in climate, to be sure. Altitude is an obvious concern. A rule of thumb says that, all other things being equal, the temperature decreases about three degrees Fahrenheit for every thousand feet in altitude. (A practical demonstration of this was noted by Dick Edwards, a meteorologist here in Albuquerque. He pointed out that while Barstow and Needles and Thermal, California, were having daily temperatures of 115°F, our daily average was 100°F, a disparity which he accounted for wholly by the difference in altitude, since the weather system was the same, covering the whole area. Albuquerque is about five thousand feet higher than Barstow.)

Similarly, the presence of large bodies of water moderates temperatures, since the water absorbs and emits heat only slowly. San Francisco, a peninsula surrounded on three sides by the ocean and the bay, is noted for its limited temperature variation. It has been observed by those who spent childhoods in Milwaukee, Wisconsin, that the lilacs bloom about a week earlier down on Cambridge Avenue, along the Milwaukee River, than they do up on Maryland Avenue, closer to Lake Michigan. This isn't very impressive data in the grand sweep of history which we are endeavoring to outline here, but it's firsthand information.

The Tibetan plateau, averaging 16,000 feet in altitude, may have a daily temperature range of a hundred degrees, from 0° F at night to 100° F at midday.

A valley on one side of a mountain ridge may be chronically wet because the air rising over the mountain cools and drops all its moisture on that side. A valley on the other side of the ridge may be chronically dry, because the moisture has all been squeezed out of the air in its passage over the ridge. A change in direction of the prevailing winds might sharply alter conditions in two such valleys.

And so on . . . local anomalies are extremely important. Societies may form in a given area because of the specially attractive climatic characteristics of that limited district. The same societies may be disbanded when those characteristics change.

A sharply defined example of this formation and dissolution of a society may be seen in the Black Sands culture in Arizona. The basic cause here was not climatic directly, but the eruption of Sunset Crater near the Grand Canyon in 1065. The volcano spread a layer of dust that held moisture and was rich in trace elements. People drifted into this area and stayed because of the improved growing conditions, forming a distinct society. Over the generations, the soil eroded away and the culture of the region disbanded with it. The time span was about three hundred years, ten generations.

In Australia there is a desert region that receives an appreciable amount of rain every year, more than ten inches. Nothing grows because the soil is deficient in trace elements, notably cobalt, which have been leached away. When cobalt is plowed into the soil, things can grow.

The same sort of problem occurs in the tropical rain forests. The rich growth of jungle vegetation fosters the common-sense illusion that the soil of the region must be very rich. However, in most areas the nutrition is found only in the thin layer of organic material which has formed on top of the soil. If that layer of material is removed, the remaining soil is very poor for growing crops. As a matter of fact, large agricultural development projects have been attempted in tropical rain-forest regions and the pattern has been very much the same in each case.

The heavy forest is cleared. The land is plowed and planted. The first crop is fair. The next is worse. The third is poor. There may not be a fourth. The "lateritic" soil in these regions literally turns to stone with its exposure to the air and the Sun in the course of ordinary agriculture. After about three years, the would-be farmers discover that they have created a large, smooth flat place in the jungle, not rich farmland. It's almost like paving the area with concrete.

When the Earth cools and times grow hard in the marginal areas, where do people come from and where do they try to go *to*? It is no surprise that, since the marginal areas do not move, the trouble rises in the same places, with the same sorts of people over and over again. They do the same things and move in the same directions.

J. S. Lee[1] pointed out in 1931 that the capital of China is moved from north to south every 800 years, and he plots three rounds of this: in the twentieth century, in the twelfth century, in the fourth century. This action in China is typical of the effects that occur when the overall climate becomes more pressing.

In this case, the nomads in the marginal areas north and east of China grow very restive and begin to nibble away at the edges of China. Their activities become increasingly troublesome to the dynasty of the time and strong efforts are made to put the nomads down. The Great Wall of China (actually a number of walls built at various times and forming a complex pattern, rather than a simple, long fence) was very serious business. The Wall was intended to keep the nomads away. Its effect was helpful, but not for long. The nomads have always invaded anyway.

At the last minute, when things really get tight, the Chinese move their capital to the south. When the barbarians finally break through, they move the capital north again and settle in to establish a new dynasty.

Historians have noticed periodic effects of this sort in the Fertile Crescent and have talked seriously about what is known in the history trade as the "millennial theory." Roughly every thousand years there is a big turmoil in the Middle East and things change. The millennial theorists mention the Semitic movement into the Fertile Crescent in the fourth millennium B.C., the Akkadian take-over of Sumer in the third millennium, the Hyksos seizure of Egypt in the second millennium, and so on.

Climatic changes have been suggested as a possible cause for this hustle and bustle, but the notion has generally been rejected on the grounds that no drastic climatic changes appear in the records. The

1. J. S. Lee, "The Periodic Recurrence of Internecine Wars in China," *China Journal of Science and Arts*, xiv, 1931:114.

difficulty that historians have faced is that they have not realized how little the climate must change to set the wheels in motion. The notion that major events occur periodically every few hundred years has recurred in the literature but has not yet been satisfactorily substantiated.

As we have mentioned, an 800-year period shows up in the tree-ring data and in migration patterns as well as Chinese history.

C. E. P. Brooks took a lot of criticism for extrapolating from known migrations to inferred climatic changes. It was sensibly pointed out that one may have migrations without climatic change and that there is no necessary cause-and-effect relationship between these things. This attitude remains valid and cannot be contradicted, yet we find that mass migrations have occurred on a fairly consistent 800-year cycle. As we try to piece together a reasonable understanding of man's activity over the last few thousand years, it is difficult to avoid the suspicion that the cycle of mass migrations is a symptom of climatic change and that we *can* work backward from migrations to other things.

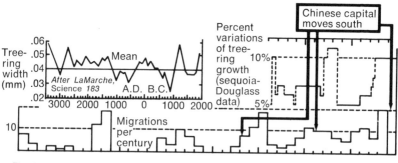

Fig. 8.5.

The graph here is a composite of things that indicate the source of our suspicions.

To begin with, we ransacked all of our sources for references to mass human migrations, which we plotted by century. As with famines, the records grow fuzzy with age, but a pattern is clearly visible.

Above the migration plot we have placed the tree-ring informa-

170

tion taken from the long-term graph of La Marche. This shows average tree-ring growth, not variability, by century.

The third line is the 1,400-year plot of California sequoia *variability,* based upon data from Douglass.

Three moves of the Chinese capital are indicated.

It is important to distinguish here between ordinary comings and goings, the adventures of armies in search of conquest and booty, and the profound uprootings that occur during the mass migrations.

Obviously, it is not possible for whole populations to pick up and move, lock, stock, and barrel, like a cloud of locusts extracting all the food from the countryside, without stirring up some conflict. Wars, battles, and tiffs accompany the migrations, as do famines and changes in dynasties.

We gain more appreciation of the significance of migration as we examine the moves themselves and learn what people do at these times.

The earliest references are to the travels across the Bering Strait land bridge about 25,000 B.C. and 9000 B.C. Our information is very scant. We are sure only that enough people moved to establish a real population in the Americas. We find traces of human habitation back to those times and certain changes in artifacts which indicate the migration of skills and knowledge carried by numbers of people.

THE EARLIEST WELL-DOCUMENTED ROUND

The picture is unclear until about 3600 B.C., when we can detect a marked change in the way of life of a settled people, the Tripolytes. These folk were farmers who lived in a region extending from the lower Danube across the Ukraine to beyond the Dnieper.

They had the standard domesticated animals, horses, cattle, sheep, goats. They spun wool, made handsome pottery, and constructed large houses with peaked roofs. They had towns with as many as two hundred houses. They worked the land, growing wheat, millet, and barley.

By 3500 B.C. these people had become nomadic. They began to move about the Steppes, abandoning a way of life that had become established over hundreds of years. One speculation has been that another nomadic people moved in on them and changed their ways, but there is no real indication of the sort of major conflict that would normally accompany a forced change of this sort. The people just

171

seem to have changed from farmers to nomads. A glance at the tree-ring plot suggests a reason. There was a peak in tree-ring growth about this time, followed by a steep decline. The Steppes are marginal lands, fertile and productive in good times, dry and unproductive in bad.

A SECOND ROUND

Around 2800 B.C. another movement is detectable. A society along the Araxes River in the Caucasus had developed a very distinctive sort of pottery. Their most unusual item was a U-shaped pottery fire dog or pot rest that was used in the hearth. The high points on this object were stylized human figures with bright, cheerful expressions.

Manufacture of these objects stopped in the area around 2800 B.C., reappearing hundreds of miles away along the Orontes River near Antioch above the ruins of burned-out settlements. Some generations later, these settlements were taken over by other people and the Araxes pottery began to appear elsewhere, in Mesopotamia and up in what is now Turkey. The Araxes hearth can be traced through several hundred years after that early migration from the region in which it developed.

A THIRD ROUND

Around 2000 B.C. an outburst of migrations occurred. We have already noted the appearance of the present great deserts of the world at about this time. Apparently the climatic change triggered a set of primary movements that carried on for centuries in a sort of domino effect.

People moved north from the Sahara region to become the Berbers.

People moved eastward from the Sahara region to settle along the Nile. The Egyptians who were already there felt this pressure.

People moved south from the Sahara.

It is noted that trade between the Indus and Mesopotamia ceased at this time.

The Indo-European peoples in southern Russia began a migration that saw them pouring through the Caucasus, then spreading east and west. These people included the Hittites, who moved into central Turkey and established the beginnings of an empire that vied with Egypt in the time of Tutankhamen. Others moved into the area of

Iran, establishing themselves as the ancestors of the later Medes and Persians.

The first Indo-Europeans pushed into Italy.

And who came to Greece at this time? The Greeks, that's who. These Indo-Europeans, pushing down from the Steppes, reached Macedonia, Thrace, and Thessaly. They are known now as the Hellenes. At the time they were probably known as "those horrible barbarians from up north."

Far to the east, at the other end of the barbarian lands, the Shang people were filtering into China and settling down. By 1750, they had become masters of the greater part of China. The Shang, also known as the Yin, were long regarded by historians as a semimythical people whose existence as a distinct entity with central power could not be demonstrated. Archeologists have now produced large amounts of documentation on the Shang.

It is known that they used chariots, a strong indicator of their origin in the Steppes, where the wheel was common. In addition, the people farmed, raised livestock, and practiced very complex technical crafts in ceramics and metallurgy. They constructed houses and other buildings in a style that is even now characteristically "Oriental" to the Western eye.

It has been noted that the Shang changed in this era from a matrilineal society to a patrilineal one. Matrilineal society is very unusual in high-latitude cultures. It is common in equatorial regions, and one may speculate (without a great deal of information to work with) that the Shang had not been high-latitude people for very long. They retained traditional ways that came with them from elsewhere.

Incidentally, rice apparently moved into China from India by way of the Yangtze around or somewhat after 2000 B.C. Rice provided heavier crop yields than any of the other cereal grains of the time and would have been an enormous advantage to people who were establishing a stable population, as the Shang were.

In the New World, at the same time, the Maya were moving into an area that they organized and held for about 3,700 years.

Farther north, in New Mexico and Arizona, the Basket Maker culture became established around 2000 B.C.

Among the secondary domino effects was the takeover of Egypt by the Hyksos about 1750 B.C. These Semitic people treated the Egyptians as subjects for about five hundred years.

173

The Hellenes kept moving, too, gradually pressing harder and harder on the Minoan civilization until they beat them decisively around 1400 B.C.[2]

Another of the domino effects that started with the 2000 B.C. activities was the eventual movement of the Aryans clear down into India, as we have noted.

A FOURTH ROUND

Apart from run-of-the-mill quarrels and more gradual movements of peoples, things quieted down until the 1300s, when mass movements started again and reached a peak in the 1200s B.C. At that time, after many decades of warm weather the climate turned sharply colder again and stayed cold until about 850 B.C.

Around 1200 B.C. a mass migration was very well documented, coming, not surprisingly, from the Caucasus. The pressure must have

2. The Hellenes had a great deal of assistance in beating the Minoans, not from other people, but from nature. It has long been known that the great Minoan city of Knossos on Crete fell about 1400. Indeed, the city was severely damaged and evidently abandoned very abruptly. It has long been speculated that raiders struck a lightning blow with such force that the population was driven off. The effect was so great that evidences of later habitation on the island of Crete show a primitive cultural development for centuries. The mystery has been why the raiders themselves did not make use of magnificent Knossos.

Within the last twenty years persistent work by archeologists has revealed that the island of Thira in the Cyclades, just north of Crete, erupted in a huge volcanic explosion about 1400 B.C., wiping out Minoan culture in a great circle. It seems likely that the whole area was visited with tidal waves, heavy dust falls, and all of the troubles that go with a major volcanic outburst in a populous area. The massive disruption to society which this entailed would have permitted the Hellenes to achieve an overbalance of power, picking up the pieces of Minoan culture.

While the eruption of Thira is technically well established, a number of historical conjectures associated with it are poorly established, but worth mention.

One of these is the thought that the eruption of Thira coincided with the exodus of the Jews from Egypt under Moses. This would place the Exodus perhaps 150 years earlier than the date generally accepted by historians now. However, an older tradition placed the Exodus in the sixteenth century B.C., associating it with the expulsion of the Hyksos from Egypt.

The middle-ground of 1400 does not conflict with other evidence (especially since there isn't very much evidence) and it does offer an explanation for two been very great, because nothing withstood it. The people in the more southerly areas, Egypt and the Levant, called the newcomers the "People of the Sea." The movement was not limited to the sea, but a remarkable fact was that when the traveling hordes reached the sea, they launched out on it, sailed to Cyprus, stomped it, then sailed along the coast.

The travelers on land moved more slowly, but they made their way irresistibly toward the south, devastating the countryside. This was not an invading army, but a swarm of peoples on the move—men, women, children, animals. They brought with them what they could, but that wasn't much. They needed to live off the land and, importantly, they wanted to find land that they could make their own.

Pushed from behind, this migrant mass swept everything before it, swelling the numbers of the group with displaced people from the lands that were overrun.

---

matters of which much is made in the Exodus story.

The first matter is the ten plagues—rivers running with blood, plagues of frogs, and so on. These effects are good descriptions of what happens when a volcano pours out a heavy flow of ash for some weeks. The area downwind is heavily coated with ash, which contributes nutrients to uncovered water. The nutrients enable the sudden growth of algae and bacteria whose reproduction is normally limited by nutrient supply. These water-dwelling organisms are often red. In this state the water is "eutrophic" and the effects are well known to biologists. If all of the water in Egypt became eutrophic under heavy clouds of ash that obscured the Sun, the frogs would be upset; disease would break out rapidly; the people would be upset—and nobody would mind much if all the slaves left town.

Whether Thira was the critical volcano or not, the ten plagues have the earmarks of heavy volcanic pollution, and whatever the true date of the Exodus, it seems likely that it coincided with major volcanic activity in the area.

The parting of the Red Sea (or, as everybody has pointed out in recent years, the "Reed Sea") could also be an effect of the volcano. Assuming that the Reed Sea was a wide tidal marshland in a shallow estuary of the Mediterranean, a tidal effect produced by the volcano might very well have drained the estuary of its customary water just as Moses got there with the Hebrews. The water might have been "out" for some hours, permitting a large body of people to cross the usually impassable estuary. When the tide returned, it would have come with a rush, as tides do, washing up much higher than usual and drowning passers-by.

It is widely suspected now, for reasons too numerous and distracting to recount here, that Thira was Atlantis.

The Hittite empire was snuffed out. The survivors of the event joined the migration and carried Hittite culture to a new home in northern Syria, maintaining their ways—but not their power—intact for several hundred more years. So completely were the Hittites obscured by this that they vanished from the consciousness of the "civilized" world for three thousand years. Though Hittites were mentioned in the Bible (Esau's mother was greatly distressed by his marriage to a Hittite woman) and occasional Egyptian records referred to them, the Hittites were assumed to be merely one tribe among the many that lived in the Middle East. Only in the last century did the remarkable size and power of the Hittite empire become clear to archeologists.

In this case we see the classic pattern of migration, conquest, settling, empire-building, and destruction. The Hittites themselves were Indo-Europeans from the Steppes who moved into Asia Minor under the pressures of climatic change around 2000 B.C. The folks they left at home struggled through, joined with others over the years, built up some strength and stability, then lost it again when they could no longer grow enough food or stand off the attacks of people even hungrier than they. The dam burst again and they followed the Hittite route to the very same place, smashing what they found.

The Sea Peoples were looking for places to settle. One group among them, the Peleset, found a place they liked near the eastern end of the Mediterranean and settled in, becoming the "Philistines" of "Palestine."

The movement of the Sea Peoples ran out of steam when it fetched up against the Egyptian armies of Ramses III, who were equipped with technology (e.g., chariots) and military techniques, which had been presented to them by the last wave of barbarian visitors, the Hyksos.

The Mycenaeans, heroes of the Homeric poems, held on for quite a while against the pressures from the north. They continued their hostilities with the people at the northwestern end of modern Turkey (apparently still independent, being west of the main route of the Sea Peoples) and brought them to a successful conclusion about 1180 B.C. That is, Troy fell to Ulysses and company.

Mycenae did not last much longer, however; barbarians from the north wore them down and plunged Greece into its "Dark Age" for four hundred years, until around 800 B.C., by which time things were warming up. This time, the horrible barbarians from up north were the Dorians. They beat up the now-settled Hellenes who had come storming in eight hundred years earlier.

At the same time the Etruscans moved into Italy and the Sicilians moved into Sicily, settling down for a good long stay.

In the same era, around 1250 B.C., the Olmecs began to settle at San Lorenzo in Mexico.

The Chavin culture was founded on the northern desert of Peru, laying the basis for the earliest real civilization in the area about 1200 B.C.

At the same time the Celts invaded England.

The Nok Iron Age culture was founded in Nigeria.

The year 1195 is the generally accepted date of the Hebrew Exodus from Egypt under Moses.

About 1165 B.C. the Hebrews attacked the Canaanites and succeeded in seizing some land, probably because the Canaanites were distracted simultaneously by the unpleasantness with the Sea People, specifically the Peleset.

Robert Silverberg has written entertainingly in *Empires in the Dust* of the proposition that the displaced Canaanites moved over to the coast and became the Phoenicians. The Phoenicians not only presented us with the alphabet, but founded colonies like Carthage, which became a great power in its own right, struggling with Rome up through the time of Hannibal.

The eleventh and twelfth centuries B.C. saw great unrest along the borders of China, and reports of the time indicate that many people with green eyes and red hair were among the troublesome nomads.

China had its usual worries, and properly, for the Chou nomadic people filtered into Shang territory, gradually taking over one major city-state after another, holding them, and settling down. At last the Shang dynasty was engulfed and the Chou dynasty appeared in its place.

Herodotus reported that around 1100 B.C. there was a drought on Thira, during which not a drop of rain fell for seven years.

A more recent study of climate in Mycenae around 1200 B.C. was done as a Ph.D. thesis in meteorology by David Lee Donley in 1971 under Reid Bryson at the University of Wisconsin. Donley established the factual existence of a Mycenaean drought pattern coincident with the decline of Mycenae. His very thorough study points out massive migrations in the area consequent to the climatic shift.

The pattern Donley documents was coincident with the end of the period of greatest coefficient of variability of tree-ring growth from Douglass's sequoia data. That is, the climate changed in California at the same time it was stirring people up in Greece.

The world had about 350 years of relative warmth and calm after the movements triggered around 1200 B.C. settled down.

By 500 B.C. the climate was turning harsh and the cycle of migrations began again, coming to a peak in the fourth century B.C.

In the interim the Hallstatt culture had become established in eastern Europe. As the climate changed, this proto-Celtic people were attacked by the Steppe nomads, including the Scythians, who were horse-mounted warriors. The pressure was great, but the Hallstatt people managed to keep the nomads to the east of the Elbe and north of the Danube. It was the beginning of the next round.

Around 400 B.C. the Chavin culture of the northern desert coast of Peru came to an end.

At the same time, the Paracas culture of the southern Peruvian desert began to expand, as did the Mochica culture.

In the fourth century B.C. there was a change in the Hallstatt culture and it was largely superseded by its outgrowth "La Tene" culture, which began a major expansion, reaching eventually from Spain to the Ukraine. They freely raided Greece and even set up a Celtic-speaking kingdom around modern Ankara in old Hittite country.

In the fourth century the Samartae crossed the Don from the east and poured into southern Russia, pushed from behind, apparently, by the Mongols.

This pressure from the Samartians on the Scythians in southern Russia gave extra impetus to the Scythian push to the west and Scythian pressure was felt along the Danube.

Rome expanded, dominating the Italian peninsula.

In 339 B.C. the Macedonians and Scythians battled along the Danube. The Scythians were defeated and their king was killed.

In 338 B.C. the Greek city-states (excluding Sparta) formed a league under Macedonian domination.

When Alexander the Great died in 323 B.C., his empire extended from India to Libya. His Macedonian armies had conquered both

Persia and Egypt after a lot of practice wrestling nomads north of Macedonia. Interestingly, Alexander's empire stopped at the Danube. He did not push into nomad territory.

Within the early part of the third century B.C. the Iron Age had begun to spread from the kingdom of Meroe in the Sudan to the west and south. There is reason to believe that ironworking was introduced into southern Africa as a result of population movements.

At the same time the Mogollon culture was being developed and expanded. It was the first sedentary culture in the southwest United States.

During the third century the Samartians dominated the entire coastal area of the Black Sea and forced the Scythians to retreat into the Crimea.

In this same time the Hiung Nu, a mixture of Mongol, Tungua, Turkic, and Finnish people, consolidated an empire that stretched from the Great Wall of China to the Caspian Sea.

Also during the third century the Toltecs moved down from the highlands to found their capital city, Teotihuacán, in the Valley of Mexico.

The Ch'in conquered all of China, having developed a tough cavalry and a militaristic government under constant pressure from the mounted nomads in the area that is now Inner Mongolia.

A SIXTH ROUND

The period from 200 B.C. to A.D. 200 was cool. The Roman Empire achieved its greatest stability in this time. Then a period of decay of stable societies occurred synchronously in Europe, Africa, Asia, and North and South America. For example, the White Dog culture in the Four Corners area of the Southwest (the Red Rock Plateau) thrived in the second and third centuries A.D. along with Rome. In the fourth century, however, the people abandoned the plateau because of drought. Later puebloans reoccupied the area.

The period of decay involved, once again, a number of mass movements of peoples as the marginal lands failed to provide what they needed to live.

Between A.D. 280 and 464 the Han Empire in China was disrupted so greatly that Chinese fled from the Yellow River basin south to the Yangtze in such numbers that the population of the Yangtze area increased by a factor of five.

Between A.D. 304 and 535 various nomadic tribes invaded and

dominated provinces in North China. It is reasonable to assume that the local inhabitants grew very tired of the whole thing.

A.D. 320 to 330: nomadic invaders destroyed the African Sudanic civilization of the Kush.

Reports are that an Ethiopian army in this same period stopped an expansion into their territory by people to the west.

The Huns and Germanic tribes moved from northern Europe and from the Steppes into southern Europe continuously between A.D. 375 and 450. The Vandals even made it all the way from somewhere around Poland down through Spain, across the strait of Gibraltar into Africa, then eastward to Carthage. From Carthage they moved up across the sea into the Mediterranean islands and took Rome from the south. This activity was very hard on the Roman Empire, which folded.

In 386 the northern dynasty of the Toba Tartars was established, lasting until 636.

The Paracas culture in the southern Peruvian desert ended about A.D. 400, eight hundred years after its thrust of expansion.

In the place of the Paracas, the Nasca-Ica culture rose.

Around 400 the Sinagua peoples migrated into the Little Colorado Valley.

The Basket Makers spread over the whole Southwest as an agricultural people about 400.

In the fourth and fifth centuries A.D. the great historical surge of Maya culture took place. Their first and greatest temple at Tikal was dated A.D. 416. There had been a great drought around 400 and the temple was useful for making the human sacrifices they found desirable in times of drought, epidemic, or invasion. During the fifth century the Maya spread down into highland Guatemala and Honduras.

From 403 to 552 the Jwen-Jwen kingdom was established from Lake Baikal north of the Great Wall to Korea.

In 420 the White Huns, originally living north of the Wall, migrated from the Jwen-Jwen–dominated area across the Oxus. From a base there, they freely raided and plundered in Persia for the next 130 years, though they never seized and held the area.

Between 500 and 550, White Huns settled and dominated northern and central India.

Between 453 and 558 the eastern Slavs migrated from the Don Valley to the Danube, the Aegean, and the Adriatic. This Slavic movement was a second domino effect, the first of the second-order reactions, and it was followed by others.

In the fifth century, northern tribes (Jutes, Picts, Irish, Frisians, Angles, and Saxons) invaded England.

British tribes fled these invaders into France, becoming the Bretons.

In 525 the Irish began settling western Scotland.

In 551 the Turks revolted against the Jwen-Jwen and drove most of them south into China, where they were absorbed. The others moved west.

In 553 the White Huns were defeated by the Turks, allowing the Persians to move north to the Oxus.

In 559 the remaining White Huns, identified by the Europeans as Avars, moved into the Russian Steppes and in just three years established an empire ranging from the Volga to the Danube to the Elbe. They subjugated the Slavs.

Between 551 and 556 the Turks expanded their empire to the Aral Sea, dominating Mongolia and large parts of Turkestan and Afghanistan.

(About A.D. 500 another volcanic event like the eruption of Thira caused an outburst of activity out of phase with the general climatically influenced events. The Tihuanaca people, who lived in a single major city near Lake Titicaca in Bolivia, were so annoyed at the destruction of the city by the eruption of nearby Cayappia that they rushed out and vigorously attacked their neighbors. Indeed, this happened again about 900, and they once more hastened out on the attack, throwing their neighbors out of phase with the climate. At length they were destroyed by the Incas and the Spanish.)

Climate and the affairs of men eased off again until nearly the twelfth century, while Europe enjoyed the Dark Ages. We should note that a series of "anomalistic" wars and migrations came to a peak in the sixth century.

While the curve of these anomalistic events built up gradually from the fourth century, it does not appear that the large number of events that occurred in the sixth century are simply domino effects—for example, the rise of Islam and its extraordinarily rapid expansion.

The entire world seems to have been affected, since there is a bulge in Chinese wars at the very same time Mohammed's people were cutting loose, and no obvious connection ties these activities together.

This puts an embarrassing bulge in our curve which we cannot readily account for. The data do not always conform to our expectations.

By the start of the twelfth century the regular cycle was back in operation.

It was noted that storms of great violence and destructiveness raged in the North and Baltic Seas. The great European rivers, including the Thames and the Po, froze over for months at a time. There were exceptional floods. Islands that had been inhabited continuously for centuries were swept dangerously by high tides, which forced their abandonment.

Norwegian harvests frequently failed. The herring ceased to spawn in the sound and migrated to the Kattegat. Open sea routes between Norway, Iceland, and Greenland were obstructed by ice, largely cutting off communication and supply to those eastern Scandinavian colonies.

China had a series a natural catastrophes—floods and droughts, pests of locusts. The Yellow River changed its course suddenly and often. The Steppe nomads harassed the Chinese steadily, disrupting orderly processes of government, transportation, and communications.

India had a drought so bad that the Mibran River disappeared.

There was a terrible drought in the southwestern United States, which is clearly documented in tree rings.

As these calamities built up to a peak, a number of major migrations inevitably took place.

Between 1096 and 1291 the Crusades carried great numbers of people from Europe to the Middle East. It seemed that everybody who could find the least excuse left Europe. They didn't just go down to take a quick swipe at the Saracens and come back, either. They established kingdoms and systematically extracted everything they could from the people who lived there, settling in to stay as long as possible.

In 1100 the Navajo and other Athabaskan Indians flowed from the north into the southwestern United States. Note that these people were on foot. The Spanish had not yet brought horses back to the New World and these people *walked* the vast distances that are today so striking to visitors in the western United States.

In 1132 the Jurchen, a people who had lived along the Amur

River, conquered and occupied all of northern China down to the Yangtze.

In 1141 the Khitan Horde, who had dominated the northern edge of China, were driven off by the Jurchen. They fled and set up the empire of Kara-Khitai.

In 1150 the Sinagua people left the Arizona highlands and moved into the river valleys.

In 1168 the Aztecs arrived in the Valley of Mexico and settled in a miserable swamp. Within 150 years they expanded through Mexico, establishing an unparalleled reputation for toughness and downright meanness.

In 1174 the Almohades, a Berber tribe, conquered and dominated Morocco and northern Africa.

In the late twelfth century, Temuchin, a Mongol whose youth was marked by persistent efforts to hunt him down and kill him, turned himself into Genghis Khan and consolidated central Asia under the Mongols.

In 1200 Mali began its expansion and by 1300 dominated the western African Sudan. Mali became in time one of the great empires of the world. When its ruler once passed through Egypt on his way to Mecca, his free spending disrupted the economy of the country.

In 1200 the Incas began to expand outward from the Peruvian high arid grasslands and conquer the surrounding area.

In the thirteenth century the Pueblo peoples withdrew from Colorado and Utah and concentrated in northern Arizona and New Mexico, especially along the Rio Grande River. Pueblo villages became more like forts. (We have talked about the fortified pueblo at the Pecos National Monument.)

Between 1212 and 1238 Spanish armies forced Moslems to retreat from southern Spain to Granada, allowing expansion of the northern Europeans toward the south.

Between 1220 and 1227 Genghis Khan conquered north China, Turkestan, Afghanistan, Persia, and southern Russia.

Between 1221 and 1239 the Kumans, a large nomadic tribe from the Steppes north of the Black Sea, migrated to Hungary, Macedonia, and Thrace.

Between 1221 and 1241 northern Russia and Korea came under Mongol control.

The Mongols held control of Russia for a couple of centuries, requiring only that the conquered people do what they were told (*exactly* what they were told, a basic tenet of Mongol beliefs being being that it was better to kill ten thousand people than let one guilty man escape punishment) and send tribute to headquarters. The Mongols had little interest in altering or participating in the way of life of their subjects. It is interesting that in his massive work *A Course in Russian History,* V. O. Klyushchevsky was able to dismiss two hundred years of association with the Mongols in a few paragraphs. Mongols weren't interested in Russians, except as a resource, and Klyushchevsky wasn't interested in them. Serves them right!

At the time, however, it seemed important.

In 1244 Karemsians fleeing before the Mongols conquered Jerusalem. They were finally peacefully absorbed by Egypt.

By 1260 all of southern China and the Middle East (with the single exception of Egypt) had fallen under Mongol control.

A long series of domino reactions followed this major sequence of pushings and pullings. For example, Tamerlane conquered vast territories from his base in Samarkand, cutting off trade between Europe and Cathay. The Europeans made concerted efforts to get around this roadblock, discovering America in the process and developing a sequence of events with which we are all comparatively familiar.

The Eskimos, regarded today almost as the last perfectly natural peaceful people, moved south in Greenland about 1400. They attacked and completely destroyed the colony of Vikings who had remained there in isolation since the ice had cut off traffic with Scandinavia. The Vikings were by then a pathetic, undernourished, undersized, rickety band, unable even with superior technology to fight off Eskimos armed with seal spears.

Seven rounds of historic migrations centered on the points of an 800-year cycle can be reasonably identified and documented. People from the same places have consistently done the same things.

The striking fact is that vastly different cultures completely out of contact with each other have gone on the move in the same way at the same time. Surely, the Incas, the Mongols, and the southwest Amerinds did not coordinate their moves. Yet they moved in synchrony.

That was about eight hundred years ago.

9

The 800-year period seems to be the major long-term cycle in the affairs of men. A longer period of 1,739 years has been suggested by Petersson, but we have not examined enough long-term data to establish that period. The information on human affairs, especially, grows rather too thin to let us trace a cycle that large back very far in time. We have no objection to the proposition, but we can't yet support it from our own investigation.

The 800-year cycle shows up in both tree rings and human affairs, and we assume from the knowledge we have of volcanic activity that these things are related. We assume further, from our knowledge of the effect of tidal forces on the processes that release energy from the crust of the Earth, that the periodic tidal force maxima and minima are the controlling factors in volcanic, therefore climatic activity. The celestial forces that govern the relative positions of Earth, Sun, and Moon cause variations in those tidal patterns. Beyond that we have not probed.

Like almost everything else, the volcanic cycle is not simple, but complex. Start with the observation that great and prolonged volcanic activity is associated with (and very probably causes) ice ages. During a period of great tidal stresses, all eligible volcanoes are triggered. On comes the ice age. The period of great stress passes. Things warm up.

Such a transition occurred at the end of the last ice age, the great ice sheets melting away by the sixth millennium B.C. The Great Climatic Optimum ensued between 5200 and 2200 B.C.

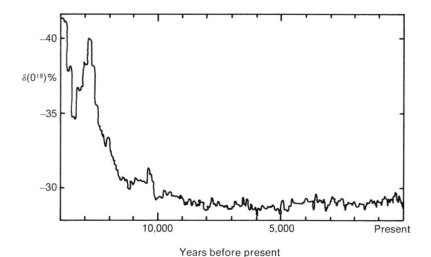

After Dansgaard et al 1971

Fig. 9.1.

If you examine the temperature plot thoughtfully, you will note that the differences between good and bad times for human beings between about 6000 B.C. and the present have been very tiny.

You have to look at Figure 9.1 very closely to spot the ups and downs with your unaided eye. To make any real sense of this period, you must expand the scale, blow the graph up appreciably, and plot more points on it so that the temperature variations that we regard as terribly important will even show to the naked eye.

All these mass movements of peoples, the rise and fall of civilizations, periods of famine and sorrow, have occurred during generally very good times. Look at the graph. The temperature shot up spectacularly about eleven thousand years ago and has stayed up. Let us raise a toast to the hardiness of our ancestors who lived through times that were more difficult than any we have seen or will see.

We have to look back 65,000 years to find a period as temperate as our own.

During those cold intervening millennia, the volcanoes gave their all. By the time the tidal forces let up, every eligible volcano must have let go with everything it had.

We observe with satisfaction that really huge volcanoes do not erupt violently very often. One is led to the reasonable suspicion that it takes a long time for a big volcano to build up the forces needed for a huge blast. Once the thing fires, it must recover for a long time before it goes again, no matter what the triggering force. Thus, the big volcano won't be affected by the tidal triggering forces until it has become "cocked" again. When the cocking energy reaches a great enough level, the additional force of the tide will be enough to trigger the volcano and fire it. Then another long cocking period must follow.

The smaller volcanoes would require less time to cock, since the forces need not build up so high. Thus, they would fire more frequently.

Consider Figure 9.2, which shows the possible cyclic effects of volcanic cocking at various rates.

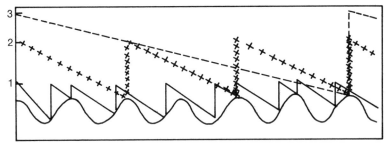

Fig. 9.2.

A theoretical cycle of tidal peak periods has been plotted along the bottom of Figure 9.2. Three volcanoes of different sizes have been indicated on the vertical scale.

Consider the smallest of these volcanoes, Number 1. Assume that it erupts at time zero on the graph. Its recovery time is not very long and its "resistance to triggering" drops fairly rapidly. Before long its resistance to triggering crosses the tidal force triggering curve and it erupts once again.

After this eruption, it is recovering at the same rate, but meanwhile the triggering forces are increasing and the resistance to triggering meets the triggering curve quite soon, bringing on another eruption. And so on.

After six tidal triggering force peaks have been reached, our Number 1 volcano has erupted ten times, never with a great deal of energy.

This little volcano is eligible for triggering a very large percentage of the time.

Number 2 also fires at time zero on the graph, but it is a much larger volcano and it takes much longer to be cocked for firing once again. After six tidal triggering rounds it has fired only four times, but its effects are appreciable.

Number 3 is a monster. After it fires at time zero it is ineligible for another eruption for a very long time. Indeed, it does not fire again until the sixth tidal triggering round.

It happens, given this choice of volcano size and recovery times, that the Number 2 and Number 3 volcanoes fire synchronously at times, and always on regular cycles, assuming a regular triggering cycle.

But look at Number 1. The times between its eruptions are quite variable, though the triggering cycle and the recovery time are both treated as constants.

It would seem that the Number 1 example is the typical case. The eruptions occur on a rational but not necessarily simple schedule. Further, the constants we have used are not actually constant. There would be significant variation in the periodicity of eruptions, though the overall effect would be generally cyclic. "Families" of volcanoes of different sizes would create dust veil cycles that do not necessarily follow a basic tidal triggering cycle in simple fashion.

Back to the original point here—after the great tidal strains of the ice age period, the big volcanoes would be exhausted. It would take a very long time for the big ones to be cocked and triggered again. It figures, then, that an ice age would be followed by comparative quiet while things got cranked up again. The very quiet Great Climatic Optimum may have been such a period.

When will the very great outbursts begin again? Nobody has done the necessary major search for such volcanic patterns.

We have calculated from the records of glacial advance and retreat, matched against the corrected carbon 14 chart, that glaciers around the world come and go on an 854-year cycle.

We can't account for this particular period. It's just a report of the news, not an editorial opinion. It does roughly fit the approximate 800-year cycle in tree rings and human affairs.

The number of glaciers in the world is surprising to most of us, who think of them as decaying relics of the past. High mountains all

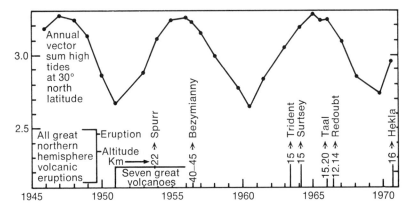

"It has long been recognized that clusters of volcanic eruptions occur periodically. The pattern is evidently continuing . . ." (From Cronin, 1971). (There were two great volcanic eruptions in the southern hemisphere: Agung, 1963, 8½° south latitude, and Fernandina, Galápagos, 1968, 0° latitude.)

Fig. 9.3.

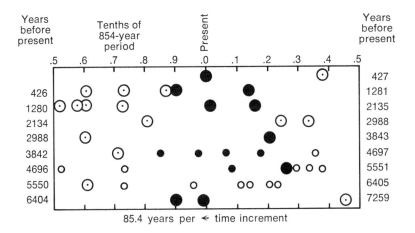

Raw data concerning glacial advances and retreats based on calendar dates that have been converted from carbon 14 dates

Fig. 9.4.

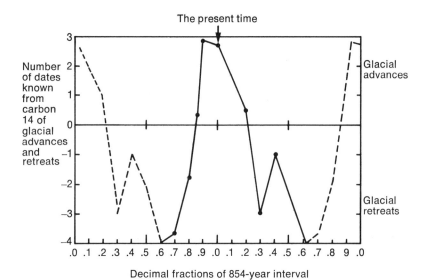

The present time

Number of dates known from carbon 14 of glacial advances and retreats

Glacial advances

Glacial retreats

Decimal fractions of 854-year interval

If a carbon 14 date is ambiguous, and could be one of four possible dates, then a value of + ¼ was assigned to the decimal fraction group to which it belonged. The (+) value is used if advance is the case; (–) if retreat.

Data after Bray 1972; also Suess

Fig. 9.5.

over the world bear glaciers. In high latitudes very large, long glaciers flow grandly through the valleys all the way to the sea. The Alps are spotted with glaciers to which people pay close attention.

At times, glaciers advance everywhere, often destroying the works of man.

Emmanuel Ladurie has identified Alpine glacial thrusts on a large scale at certain times, with additional periods of smaller oscillation between periods of glacial retreat.

The Alpine glacial thrusts maximized between:

1400 and 1300 B.C. (greatest)

900–300 B.C. (only a century and a half of withdrawal in this time—the Subatlantic Era)

400–750 A.D. (after a glacial retreat corresponding to the time span of the Roman era)

1200–1300 A.D. (the "Little Ice Age," with high average glaciation with temporary oscillations)

About A.D. 1200 a 200-year-old Alpine forest was killed by advancing glaciers.

Yes, the glacial data fit the pattern.

Other things fit, too.

C. G. Abbot has a great deal to say in favor of a forty-five-year cycle that he believes is very significant, marking sorry events like the Great Depression of the 1930s, the financial and agricultural problems of 1886, and the depression of 1838.

Abbot's whole rationale is based on sunspot cycles, of which we have spoken only briefly. The existence of the cycles is not seriously disputed now, after centuries of observation, but the *effect* of sunspots upon climate is not so readily accepted.

Still, Abbot makes a good case, correlating precipitation records, droughts, the rise and fall of levels in lakes at various locations.

Abbot is concerned chiefly with a "double double" sunspot cycle —22.75 years taken twice.

Our study suggests that the period is right, but the factors are more complex, including not only the sunspot cycles, but ten rounds of perigee precession, five rounds of eclipse node semicycle, and the yearly perihelion cycle.

Forty-five-year cycles show up in the tree rings, according to Douglass, and Abbot insists that drought is the keynote of the time. Using Krick's computer data, Harry Geise has pointed to drought coinciding with Abbot's 45-year cycle schedule . . . due in 1975.

Of course, the effect of this cycle depends a great deal on the configuration of the 800-year curve at the same time.

800-year basic cycle
45-year "grass" riding
on basic curve

Fig. 9.6.

If the 45-year peak is added to the *top* of an 800-year peak, the effect of the 800-year peak is increased. If the 45-year peak occurs in a *trough* of the 800-year cycle, the 45-year peak moderates the effect of that trough and is itself not very important.

The double sunspot cycle, at 22.75 years, does seem to correlate closely with precipitation. The increased number of sunspots comes at the same time as increased rainfall, and at the same time as bad wine harvests, and at the same time as Republican administrations in Washington.

We must concede at once that these are merely correlations, that we cannot prove satisfactorily that Republican political success is caused by sunspots. It may be that the Republicans blight the wine and cause the sunspots or that poor wine harvests depress the opponents of the Republicans and the Sun breaks out in sympathetic

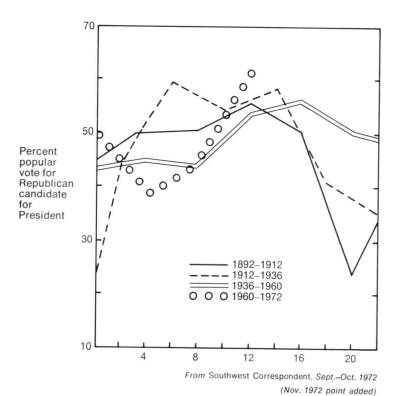

From Southwest Correspondent, *Sept.–Oct. 1972*
(*Nov. 1972 point added*)

Fig. 9.7.   Years in a voting cycle

hives. It may all be coincidence. Nevertheless, we have our own suspicions in the matter.

Figure 9.7 plots the percentage of votes for the Republican presidential candidate in elections between 1892 and 1972.

The plot is shown against a twenty-two-year field, not quite in phase with the sunspot cycle. The peak of this wave form will gradually precess to the right as we add data, but these four rounds are not appreciably altered by the error. The curve is remarkably consistent. Why plot Republicans instead of Democrats? They are not quite reciprocal. Apparently, "conservatives" are less likely to split into factions than "liberals" and are therefore easier to trace over the long period shown. The "conservatives" split only when Teddy Roosevelt ran off with the Bull Moose and then when George Wallace took 13 percent of the vote. The liberals quarreled among themselves rather more often.

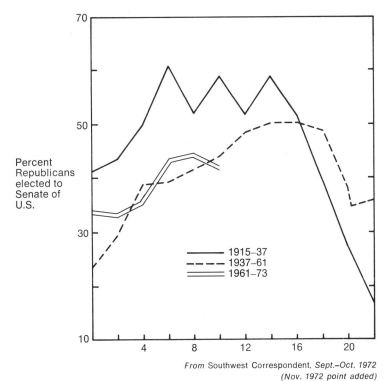

Percent Republicans elected to Senate of U.S.

1915–37
1937–61
1961–73

From Southwest Correspondent, *Sept.–Oct. 1972*
(Nov. 1972 point added)

Fig. 9.8.   Years in voting cycle

The pattern holds for the Senate as well as the presidency, as Figure 9.8 shows.

The single sunspot cycle, rounded off at 11.4 years, is associated with a series of social events that may easily be declared coincidental, but which are interesting anyway. Taking into account the fact that there is about a two-year lag after the sunspot maximum until the break in the rate of rainfall, the following is the schedule of events of the last hundred years.

This presentation of events may not be rationally compelling, since the list of events is not comprehensive. Yet the presentation is unsettling, since it does touch high points with fascinating consistency.

The Russian Virgin Lands program is a sensible large-scale effort to turn vast areas of fallow land to agriculture on a crash basis. Massive purchases of wheat by Russia in recent years suggest that the program is not outstandingly successful.

The program is operating largely in the Steppes, the place that the nomads traditionally settle and then abandon in difficult times. Indeed, one of the places where the Russians are most assiduously endeavoring to produce major food crops is called the "Famine Steppe." Little hints like this give a body pause to wonder if the plans were really well laid.

We have another body of information that is very shaky in scientific terms, but too interesting to ignore. The fact is that people have written down a lot of comments about events over the centuries. The form of the writings is often baffling and the interests of the writers do not always coincide with the interests of the modern reader. However, some of the material is both entertaining and rich in data. Herodotus, writing around 450 B.C. has given us great quantities of useful information about places and people. In recent years Herodotus has been confirmed on a number of points which had previously been shrugged off as errors. His descriptions of the Scythians, of African Pygmies, of the reported watercourse from Lake Chad to the Nile, all contribute fascinating circumstantial detail to the times about which he wrote.

The Nihongi Chronicles of Japan from the Earliest Times to A.D. 697 is a remarkable document containing both dreary detail about who was granted what titles when and entertaining gossip about spicy scandals at court.

The Nihongi provides a wealth of useful detail for our purposes—

194

Table 9.1.

United States		U.S.S.R.
	1849.6	Revolutions of 1848; persecution of intellectuals. Marx's *Communist Manifesto.* * Emancipation of serfs to avoid revolution. Began conquest of Asia.
	+ 11.4	
Civil War begins	1861.0	
	+ 11.4	
Panic of 1873	1872.4	
	+ 11.4	
Revolutionary Congress in Pittsburgh in 1883. Publications on making explosives and Molotov cocktails. Panic of 1884.	1883.8	
	+ 11.4	
	1895.2	Beginning of concerted Revolutionary struggle, 1896. First strike of 30,000 in St. Petersburg.
	+ 11.4	
Financial Panic of 1907	1906.6	Russian Revolution (1905). Russo-Japanese War. Pogroms initiated.
	+ 11.4	
World War I followed by depression of 1921	1918.0	World War I followed by Bolshevik Revolution.
	+ 11.4	
Beginning of drought and Great Depression	1929.4	Beginning of drought, depression, and liquidation of millions. Introduction of compulsory collectivization.
	+ 11.4	
World War II	1940.8	World War II. Rebellion in Ukraine.
	+ 11.4	
Recession; drought; Korean War	1952.2	Massive crop losses. Beginning of Virgin Lands program. Death of Stalin, execution of Beria. East Germany revolted. Malenkov deposed in 1954.
	+ 11.4	
Viet Nam War	1963.6	Crop failures—purchased 8 million metric tons of wheat. Khrushchev deposed in 1964.
	+ 11.4	
?	1975.0	?

*At the same time in 1848 there was revolution in France and the Second Republic was set up. In China the Taiping Rebellion was reaching the bursting point, at which it produced unpleasant effects already noted.

reports of unusual weather (summer snows, premature blossoming of the peach trees, followed by frost, etc.), crop failures, uprisings, epidemics, movements of the capital, and, very significantly, reports of eclipses and volcanic eruptions.

It is common to find sequences of events, day by day, like these:

Strange mists.

A great storm.

First appearance of peach blossoms in the spring.

Destruction of plants by hail.

Uncommon wind, thunder, rain, and sleet, with alterations in the laws to accommodate the unusual events.

A great fire.

Flowers and herbs damaged by frost.

A great rainstorm.

Unusual cold and wind.

Hail—people dressed in heavy clothes.

Reports to the Empress of hard times in the provinces.

Report of hail an inch in diameter.

Removal of the residence of the Empress to a temporary palace.

An eclipse of the moon.

"In this month the water of the Mamuta pond changed and became like indigo juice. Its surface was covered with dead grubs. . . . Water in the drains coagulated to a thickness of three or four inches. . . . Fishes stank and were unfit for food. . . ." This is a good description of eutrophic waters, probably caused by an infusion of volcanic ash.

The Nihongi reports frequent dissatisfaction with omens like thunder on a clear day. In a land of volcanoes, thunder on a clear day is likely to mean bad news.

Thucydides and Herodotus both report sequences of events like those in the Nihongi, repeatedly linking earthquakes, volcanic eruptions, eclipses, famines, epidemics, and unhappy times. Earthquakes and eclipses are not just miscellaneous troubles lumped with everything else on general principle, but appear to be specifically related to other events in the historical literature.

Surely the material for dozens of graduate student theses is buried in these works. Years will be required to ferret out references in European, Chinese, Japanese, Indian, and South American sources, and plot them, so that we may see what correlations actually appear in the diverse literature.

We have examined the English chronicles, in which one of the items they conscientiously observed and noted was the occurrence of "unusual climatic conditions." Between 1000 and 1450 these records were maintained consistently enough so that we can reasonably plot the data to reveal patterns. We did plot the frequency of reports of unusual climatic conditions, by century, in that period (Figure 9.9).

The reported occurrence of unusual climatic conditions in that period is clearly not random. The reports build to a high peak in the 1300s, not surprisingly.

Now, here comes the shaky part.

Since we didn't have chronicles for other periods, we tried a method for constructing records equivalent to the reports we *did* have in the one 450-year period.

We examined the astronomical records and determined the actual alignments of tidal triggering forces during that 450 years—eclipses and coincidence of eclipses with perihelion. These were summarized by century.

Then we examined the calculations for every century from 700 B.C. to A.D. 2200 and determined what those same triggering forces were—

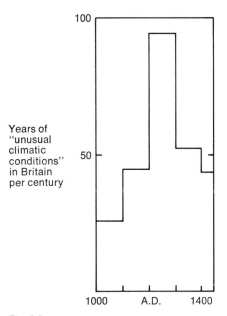

Fig. 9.9.

counting eclipses and perihelion coincidences.

Then from those cumulative triggering forces in each century we calculated what the "unusual climatic conditions" would have been *if the same relationships between tidal forces and climate applied in*

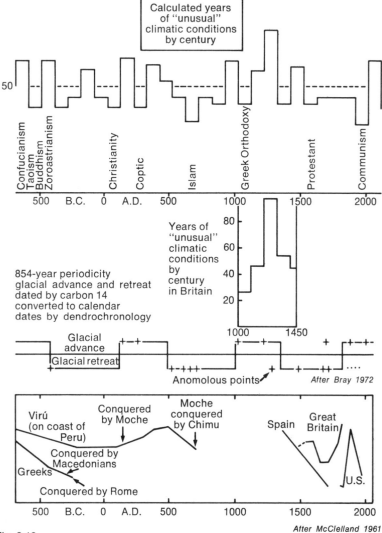

Fig. 9.10.

all the unknown centuries as actually occurred in the reported centuries.

This created Figure 9.10.

This treats only Britain (above 50° north latitude), and takes into account only a couple of factors. This is the rankest sort of awkward extrapolation, making a mountain of a molehill of data and a handful of conjecture.

But—the product does check out rather well against other measures.

For example, the comparatively warm and quiet era we have enjoyed in the first half of this century is distinctly indicated by these calculations.

Notice that the times at which the great religions of the world were formed were all comparatively quiet eras when people had an opportunity to be contemplative and philosophical without necessarily risking their lives.

When the glacial advances and retreats are plotted against this chart, the fit is not bad.

In addition, we plotted available pieces of behavioral data taken from McClelland on the same time base. When the "need of achievement" indicators in a society *in*crease in the areas indicated, it is *in general* a sign that the average temperature is *de*creasing. The indicators follow glacial advances and retreats remarkably well, and thus also follow our half-baked extrapolation rather well.

By any test we have been able to apply, history and climate have correlated closely.

But enough of history.

Where are we now?

We are once again in a time of great change. We have reached one of those harrowing periods when people are on the move. It must be kept in mind that the outstanding characteristic of these times is not simply the average lower temperature, but great climatic variability. While the climate patterns we have identified are *average* conditions, the extremes tend to be much higher and much lower. For example, when things grew cool around A.D. 1200, the average precipitation was higher in the 20°–40° band. However, the Southwest experienced very severe drought within that period also. Drought did not predominate, but its effect was devastating at times. The extremes stir people to action. Climatic extremes produce mass movements.

199

Small deviations in averages produce trends that are slow enough to be assimilated.

The 800-year period is here again.

The 45-year period is here again.

The 11.4-year period is almost here again.

This decade and century are getting the excitement we should expect if there is anything to this whole matter we have been discussing. Consider some indicators.

The glaciers are on the advance all over the world. Russia reported in 1973 that two Siberian villages had been abandoned because of advancing glacial ice. Icebergs have been reported in the shipping lanes of the North Atlantic for the first time in many years and the number of reports is increasing sharply. We believed that radar and other modern technology had made icebergs nonnewsworthy, but that was an illusion. Now that the icebergs are returning, we hear more about them.

Meteorologists report that the average world temperature has dropped approximately 2.7° F since 1950.

There are indicators that large areas of the world at latitudes above the 45° crossover point are growing drier and colder in accordance with the typical patterns of the past. The northwestern United States has been having drought conditions. Indications are that Russia's crop shortage problems are associated with drought.

Significantly, the massive hydroelectric system of the northern United States has been affected by the lack of rain. The Columbia River basin, for example, has been so short of water that the power plants of New Mexico have been supplying some electricity to the Northwest.

Canada, too, has an enormous investment in and dependence upon the hydroelectric facilities near its Pacific coast. One begins to feel queasy about all this with the realization that a huge percentage of North American facilities for production of electricity-dependent aluminum have been constructed in this area over the last quarter of a century. While environmentalists have been infuriated by the appearance of smog in the sparsely populated Four Corners area, which is now supplying some of that electricity, they may find themselves having to choose between smog and aluminum.

Analysis of satellite pictures shows that the Earth's albedo is increasing distinctly. That is, snows on the ground are reflecting solar radiation so that less heat is absorbed. This cools the Earth still more, making for more snow, which stays on the ground to increase albedo.

These are little things, perhaps, straws in the wind.

Scandinavians in Greenland who have studied their own history may be forgiven if they keep a watchful eye on the Eskimos.

Americans think of Oklahoma as the center of the Dust Bowl of the thirties, but the reference books based on meteorological data point to the Dakotas as the area hardest hit by drought, slightly farther north.

The capital of China moved south to Chungking when the Chinese Kuomintang gained control of the government from the last of the decadent Manchu rulers in the early part of this century. This move was right on schedule.

China was torn by strife in the period following the fall of the Manchu dynasty and many different factions wished to rule. Mao Tse-tung was among the contestants. He and his small band of supporters tried without success to raise a significant army in south China. They simply couldn't get enough support in China to give them power.

Mao and his army were forced into the famous Long March of the thirties, during which they were cut to ribbons as they walked all the way up to the edge of the Ordos desert at the Great Bend of the Yellow River, close to the Great Wall. Mao made his headquarters there for many years, consolidating his position and winning a lot of points by joining the fight against the Japanese invaders during World War II.

After the war, Mao turned back to serious business. He did something out there at the edge of the desert which he had been unable to do in south China. He raised a significant army of rough, tough, hungry people who responded with enthusiasm to his suggestion that they march down from this marginal area and seize China. Under the slogan of "land reform," Mao's army swept in from nomad-land.

With their power established, they moved the capital back to Peking in the north.

Having followed the route themselves, Mao and his associates are distinctly aware that bad news comes to China repeatedly from Mongolia, Manchuria, Turkestan, Dzungaria. Steps have been taken to protect China against invasion from these areas.

China has been expanded—into Tibet, for example. The borders are tested constantly. China has fought skirmishes with India over disputed border areas in places that hardly seem worth talking about, let alone fighting over.

China sent troops into Korea during our own hassle there.

The Chinese-Russian border along the Amur River to the north of China is the scene of differences of opinion which occasionally escalate into full-fledged tank and artillery duels, with rushes of infantry across the frozen river. This acrimonious debate is surely not just whimsical. While it is not unusual for people to jest rudely with their rivals and competitors, tank battles betray concern that goes beyond humor. The People's Republic of China takes profound interest in the same borders that the Shang and the Chou and the Han dynasties worried about in millennia past.

Sensibly, Mao has settled perhaps fifty million Chinese in those marginal borderlands to cultivate the soil, establishing a buffer population between China proper and the barbarians.

As we have noted, Russia has transported large numbers of people into its borderlands, the Steppes of Central Asia, not very distant from China. The people they sent to the Famine Steppe and environs were largely Lithuanians, Latvians, Estonians, East Germans, Poles, and others who may not have been too crazy about going, but who yielded to persuasion. Thus, Russia established a good-sized buffer population between itself and the place where the barbarians have traditionally come from.

Both China and Russia have been buying food in large quantity from the United States, Canada, and Australia. One wonders about the marginal lands of Central Asia, where eighty million people are in a deteriorating situation. Their predecessors have traditionally moved out, south through the Caucasus, westward into Europe, eastward into China. Perhaps these people are more philosophical and are simply accepting with a rueful smile the increasing probability of starvation.

In North Africa a drought has persisted for several years and the unhappy effects are becoming apparent to outsiders in increasing detail. The area of concern is the "Sahelian zone," the arid country along the southern edge of the Sahara desert. In good times this land, unlike the Sahara proper, supports many nomadic peoples with large numbers of livestock. As one travels south through the region, the vegetation shifts gradually to wetter, heavier growth and eventually changes to the jungles of the equatorial region.

The Sahelian lands are Mauretania, Senegal, Mali, Upper Volta, Niger, and Chad. Large areas of the Sudan and Ethiopia lie at the same latitudes as the Sahelian zone and climatic effects there are similar.

The Sahelian drought has allowed the Sahara to expand as far as a hundred miles to the south. It has been calculated that the 0.5° C average temperature drop that has occurred in the Arctic would reduce rainfall in northern Nigeria by ten to fourteen inches. Perhaps this is what the Biafran war was about.

Famine is sweeping this vast region with such ferocity that whole peoples may vanish from the Earth. The fascinating desert tribesmen whose pictures have adorned the pages of the *National Geographic* may soon be just memories, since the people themselves will have died or scattered, losing their tribal identity.

By 1973 as much as 80 percent of the livestock in wide areas had died of thirst and hunger or had been slaughtered for food. There were no crops. The people had—and have—no hope. Nothing grows. Nothing lives. Recovery is not possible. When the area at last grows wetter and warmer again, population will reenter it from outside. The old tribes are leaving no seed.

As this disaster has advanced, refugees—nomadic peoples without alternative—have streamed to the south. The Sahara blocks northern migration, so the Arab lands along the Mediterranean have not been nearly as hard pressed as the Central and West African nations.

Herders have brought their livestock into settled areas where small amounts of food may still grow. Cattleman/sodbuster wars have broken out in Upper Volta. The refugees are swelling the populations of countries like Nigeria and straining the resources of those countries.

Reports are that basic crops like peanuts have been reduced by 30 to 50 percent in Nigeria itself.

Lake Chad is so reduced in size that a lakeside fishing town is reported to be eighteen miles from water. The lake has broken into four separate bodies of water.

People are dying here and there and everywhere—hundreds of thousands, millions. The end is not in sight.

It is felt by many people in the area that mismanagement of economics and ritual is at the root of the problem. Efforts are being made to change the management. Niger, wholly within the Sahelian zone, has had a coup, as have several other countries. In Ethiopia even Emperor Haile Selassie has been undone. From his point of view, the drought has some compensating virtues. For one thing, it is killing many Moslems in northern Ethiopia (Eritrea). The central government of Ethiopia is Coptic Christian—an ancient schismatic form of Christianity which has gripped Ethiopia firmly for many hundreds of years. The Coptic peoples and the Moslems have never quite seen eye to eye on the appropriate nature of the state religion and the choice of leadership of the government. The Ethiopian army has endeavored for decades to resolve these differences by reducing the numbers of contentious Moslems. This has proved difficult, but the drought is solving the problem. There may soon be no argument about control of Eritrea.

Reid A. Bryson (chairman of the Department of Meteorology at the University of Wisconsin and the articulate chief proponent in the United States of the notion that world climate is really changing, not just passing through a freakish temporary phase) has discussed the "Sahelian effect" in considerable technical detail. He has accounted in meteorological terms for the extended drought in the region.

For example, it is Bryson who points out the effect that the location of the circumpolar vortex (discussed in Chapter Five) has on the monsoons upon which the Sahelian and other areas are utterly dependent. A decrease in polar temperature keeps the edge of the circumpolar vortex closer to the equator, blocking the monsoon from the places where it is critically necessary. The temperature has gone down. The monsoons *are* failing. Bryson sees this as a persistent situation. There is no joy in Mudville.

Another region wholly dependent on the monsoons is India. When the monsoons are blocked, India's central and southern regions parch.

India has the same problem as other areas, like the Sahelian zone, but somehow India manages to magnify the problems almost beyond comprehension. Its outstanding problem is its dense, unwieldy population, which defeats all attempts at efficiency in agriculture and distribution. Things have always been edgy in India, but its situation at the moment is especially strained. India has had comparatively good times over the last sixty years.

When the volcanoes quieted down after the first two decades of this century, the frequency of droughts in India declined sharply. The number of famines in India also declined (a million died here, a hundred thousand there, but average was modest by local standards), and by the 1960s when the Green Revolution began to increase crop yields, Indians announced the expectations that problems with hunger and poverty were ending.

An Indian diplomat once observed that while baseball is the national sport of the United States, sex is the national sport of India. With an eye on the good times, India happily let its population boom at the rate of about 2.5 percent a year, demonstrating once more Parkinson's law: "Expenditure rises to meet income."

A recent estimate says that 560 million people inhabit India. The country would not be significantly depopulated if 10 percent of its inhabitants starved to death—but that modest decrease would involve the deaths of 56 million people.

The volcanoes have started up again. The temperature is dropping. The Green Revolution is fizzling in countries that cannot provide the heavy capital required to make it work. The monsoons are failing in India.

The monsoon failed at the start of the 1972–73 crop-growing season. The resulting crop was a staggering 40 percent lower than planned, and India exhausted its resources purchasing food from abroad. Between mid-1972 and early 1974 the cost of a ton of wheat delivered to the dock in India rose from about $80 to about $250.

The monsoon came through decently for the 1973–74 growing season, but the lesser winter rains that are needed to maintain the crops were very light. At this writing, the 1974 crop prospects were looking grim. A real shortfall could push India over the edge into catastrophe. The United States has traditionally helped India by sending surplus foods—*$9 billion* worth over a couple of decades.

The United States no longer has surpluses.

The Philippines, Japan, Indonesia, along with India, Russia, and China, are major food importers.

We are on the ragged edge of something very unpleasant.

What about migrations? Are large populations on the move?

Yes, on a scale that is comparable to the mass movements of the past, and still building. Since 1933 . . .

> 33,000,000 were displaced within prewar boundaries of their own countries.
>
> 7,000,000 went to foreign countries.
>
> 1,500,000 were gathered into camps in Germany, Austria, Italy.
>
> 500,000 fled from Catalonia.

58,683 Estonians were refugees from the U.S.S.R.

153,000 Latvians fled the U.S.S.R.

78,775 Poles fled the influence of the U.S.S.R.

348,000 Ukrainians fled the U.S.S.R.

90,000 Jews fled the U.S.S.R.

35,000,000 Chinese were uprooted by the Japanese in World War II. Of these, 3,500,000 were left homeless and 1,750,-000 settled in Hong Kong and Macao.

1,200,000 Arabs were displaced by the Arab-Israeli wars, and of these, about 30,000 are being resettled per year. It's a losing proposition, because 42,000 children are being born in the refugee camps each year—partly because pregnant women are provided with larger food rations, in effect subsidizing the birth rate.

14,000,000 were displaced by the Indian-Pakistani partition in 1948.

1,500,000 refugees passed through Austria following the Second World War: Germans, Hungarians, Yugoslavs, who passed through to France, Belgium, Norway, Sweden, Switzerland, and the United Kingdom.

14,500,000 overseas Chinese are in Southeast Asia.

From Communist countries there has been an almost constant flow.

5,500,000 were displaced by the partition of Korea.

3,500,000 fled from East Germany. About 25 percent of those were repatriated.

20,000 Europeans fled from China.

53,000 fled from Yugoslavia.

200,000 fled Hungary.

350,000 fled Cuba.

1,000,000 fled from North to South Viet Nam.

12,000,000 Germans were expelled from eastern Europe.

1,000,000 South Vietnamese fled the countryside to government-controlled areas.

2,500,000 Jews have gone to the Middle East. This is unusual in being a gathering of peoples, rather than a dispersal in flight.

From Europe, just between 1952 and 1958 . . .

385,000 refugees fled Europe, in addition to

570,000 who simply chose to emigrate.

In Africa . . .

250,000 fled from Angola, Sudan, and Rwanda to the Congo (Zaïre).

50,000 Rwandese fled to Burundi.

30,000 fled Rwanda and Mozambique to Tanzania.

100,000 Rwandese and Sudanese fled to Uganda.

3,000 Rwandese not settled elsewhere were accepted by Tanzania.

60,000 Congolese fled to Uganda and Burundi.

200,000 Algerians fled to Tunisia and Morocco, but were ultimately repatriated.

There were other odd-lot migrations:

12,700 White Russians fled from mainland China through Hong Kong and dispersed.

5,000 Indonesians were resettled overseas from the Netherlands.

750,000 French and Algerians fled Algeria.

15,000,000 U.S. citizens have moved in two primary ways. General population moved from noncoastal to coastal states. Blacks moved from the country to the city. These trends seem just now to be reversing. There is migration *to* Nebraska, for example.

6,000,000 immigrants not listed above have gathered in the United States.

And this is just up through 1970. Much of the world's population has moved since then and the trends are shifting significantly (e.g., Americans are moving away from coastal states again), but the list grows unwieldly.

In somewhat less than forty years, about 3 percent of mankind has migrated—not just moving across town, but abandoning cultural roots and setting out into the great unknown.

Within the 1970s, the migrations have continued at an increasing pace. For example:

Ten million people fled East Pakistan and went to India, largely because there was insufficient food in East Pakistan and everybody was growing mean and cross. The problems were exacerbated by the appearance of a great typhoon in the Bay of Bengal which caused terrible tidal waves and floods, killing more than 200,000 people in a brief time, spreading disease, and destroying food crops.

Anybody who goes to India in search of an improved standard of living has a special point of view on things. India detected a potential problem in this migration and encouraged the refugees to go home as soon as possibile. The return to home was impeded by strained relations between the East Pakistanis and the representatives of the Pakistani government, centered in West Pakistan, far away across India. The strain took the form of riot, revolution, and killing. To speed the parting guests, India intervened, warred vigorously against the Pakistanis, and beat them.

Happily proclaiming the independence of East Pakistan (which changed its name to Bangladesh), India sent the visitors home in triumph to find their own food.

(Not that it has any discernible connection with climate, but it is interesting to note that the three countries that have enjoyed the tenure of women as their prime ministers in recent years have used government military power with great frequency. Indira Gandhi (daughter of Jawaharlal Nehru, who was a close associate of the apostle of non-violence, sainted Gandhiji) has warred with China, Portugal, and Pakistan very effectively. Mrs. Bandaranaike of Sri Lanka has suppressed student uprisings with firmness. Reports suggest that as many as 50,000 people have been killed in recent years by government troops. Golda Meir presided over differences of opinion with her neighbors which have produced guerrilla activity, commando raids, and all-out wars.)

It is somewhat difficult to obtain accurate information on the number of war deaths and war-related deaths as in epidemics, but a rough cut at the numbers in the past century goes like this:

10,000,000 dead directly from World War I.

21,640,000 dead from war-related influenza between May and November 1918.

40,000,000 dead from World War II.

500,000 dead in sequel to Bolshevik Revolution, 1918–20.

1,250,000 in first Chinese Communist War, 1927–36.

2,000,000 in Spanish Civil War, 1936–39.

5,000,000 in the partition of India/Pakistan, 1946 et sequitur.

1,000,000 Biafrans (Ibos) vs. Nigerians.

94,800,000 executed by the Communists.

250,000 in the Franco-Prussian War, 1870–71.

250,000 in a Moslem rebellion in west China, 1861–78.

200,000 in the Cuban Revolt, 1868–78.

250,000 in the Russo-Turkish War, 1877–78.

200,000 in the Spanish-American War, 1898.

160,000 in the Colombian Civil War, 1899–1902.

300,000 in the Colombian Civil War, 1952–54.

250,000 the Maji-Maji Rebellion, 1905.

250,000 in war in Achin, 1873–1908.

250,000 in the Mexican Revolution, 1910–20.

200,000 in a Moslem rebellion in Kansu, China, 1928.

400,000 in a war between the Sudanese Moslems and animists, 1960–70.

This affair was settled with Haile Selassie, of all people, as a peacemaker, with an arrangement which has caused surprisingly little comment in the West. The women and children of the black animists are detained inside their villages all of the time, while the men are required to go out and work to support them. The Moslems guard the villages and take what they consider a reasonable toll from the men returning to their families in the villages. This area must by now be much affected by the drought and it is hard to imagine that the system can be maintained, but it must have seemed like the thing to do at the time.

200,000 in the Congolese War, 1960–70.

2,000,000 estimated casualties in some 300 smaller wars.

This accounts for something over 180,000,000 deaths produced directly or indirectly by wars. A drop in the bucket, you say, compared with the total population of the world? Maybe so, but it does seem like a sign of the times.

Most of these were killed in the attempt of Europeans to expand or of Asiatics to expand either west or south.

One is especially struck by the parallels of the establishment of modern Israel with the earlier Crusades. Declaring that they had long ago left something in the Middle East which they now wanted back, people have poured into the area both times, chiefly from Europe. The effort is to settle and make a home, not just loot and leave. During the Crusades, people were leaving terrible places up north for something that might be better. Some were disillusioned and left, others stayed until they were driven out a couple of hundred years later. Descendants of others are still there. This does not sound wholly unlike modern Israel.

Table 9.2.	Communist (20th Century)	Mongolian (12th Century)	Steppe Nomads (4th Century)
Afganistan		(X)	(X)
Albania	(X)		(X)
Austria	P X	P X	(X)
Belgium			P (X)
Bulgaria	(X)	P (X)	P (X)
Chile	(X) *2		
China (North)	(X)	(X)	(X)
China (South)	(X)	(X)	
Cuba	(X)		
Czechoslovakia	(X)	X	(X)
Egypt		A	
Germany (East)	(X)	P X	(X)
Germany (West)			(X)
Greece	A		(X)
Hungary	(X)	X	(X)
India (North)	P (X)	P (X)	(X)
India (South)	P (X) *2		
Iran	P (X)	(X)	X
Iraq		(X)	X
Israel		P (X)	
Japan	P (X)	A	
Jordan		(X)	
Korea (North)	(X)	(X)	(X)
Korea (South)	A	(X)	(X)
Laos	P (X)	P (X)	
Lebanon		(X)	
Luxembourg			(X)
Mongolian People's Rep.	(X)	(X)	(X)
Netherlands			P (X)
Pakistan (West)	P (X)	X	(X)
Poland	(X)	X	(X)
Romania	(X)	P (X)	(X)
San Marino	(X) *2		
Syria	P (X) *2	(X)	
Tibet	(X)	(X)	(X)
Turkey		P (X)	P (X)
U.S.S.R.	(X)	(X)	(X)
Vietnam (North)	(X)	(X)	
Vietnam (South)	A		
Yugoslavia	(X)	X	P (X)

X Conquered   (X) Conquered and occupied   P Partially
A Attacked unsuccessfully   *1 Elected and still in power   *2 Elected and now out of power

In the superb work *The Statistics of Deadly Quarrels*,[1] Richardson classified wars by number of deaths involved, "plus or minus 50 percent". The smallest quarrel to count as a war kills a thousand people, plus or minus 50 percent.

By this reckoning, the quarrel in Northern Ireland has become a real war.

All in all, in the last century of war and forty years of estimated displacements, about one person in seventeen has been killed or forcibly displaced in the entire world, or has died of directly associated causes.

The people who control the Steppes have repeatedly come boiling out of the area and seized or tried to seize the same territories. We are in a period of such activity right now. Table 9.2 shows the effect of the last three rounds.

Importantly, we are probably just in the early stages of this activity. Even if the climate shifted suddenly back to the patterns and averages of the 1950s, the dominoes that must fall from the problems of the last few years could not be stopped.

There are *no* obvious indications that the climate will become anything but colder and more variable over the next five decades.

1. L. F. Richardson, *The Statistics of Deadly Quarrels* (New York: Quadrangle Books, 1960).

10

History may be entertaining, but the future is serious business. We strive in high-latitude countries to anticipate what is coming as accurately as possible.

What *is* coming?

We can answer only in general terms. If our perception of the relationship between tidal forces and climate is correct, then we expect the average temperature of the Earth to keep going down, perhaps another 1 to 1.5° F over the next fifty-five years. We'd expect the climate then to come back up to where we are now in another 110 years. This course of events could be altered, if not made more cheerful, by an asteroid strike on Earth or Moon.

Bryson asked and answered a question with respect to the Sahelian drought: "Will the monsoons return? Probably not regularly in this century."

The climate will continue to be variable, with new highs, new lows, new precipitation records, droughts, big storms, and high tides. From the farmer's point of view, this is not heartwarming news. It takes a great deal of confidence in the repeatability of miracles to sprinkle seeds in the ground and expect to harvest crops months later. The confidence is based largely on the belief that conditions are more likely to be stable than not. Farmers don't like surprises. It appears that we will be dealing with a constant stream of surprises, but we hope that we can spoil some of them by learning to anticipate them. Forewarned is forearmed, and all that.

We have no mystical fascination with cycles, per se. We do not see any great moral virtue in cycles or preach a particular ethical viewpoint involving cycles. Indeed, the periods we detect in the time spans that can be examined are but passing humors of the universe. The length of the year changes. The length of the sunspot cycle changes. The length of the day changes. It just happens that, compared with the length of time that human beings have spent on Earth, the period in which these cycles have been almost constant is very long. Everything that is familiar will pass, but in these few thousands of years that interest us, things have been occurring on a consistent schedule. In a billion years nobody will care, but late in the twentieth century we have something at stake.

The next two or three years—1975, 1976, and perhaps 1977—should provide relief to the planners and organizers around the world. If Abbot's 45-year cycle effect comes through, then the monsoons should return to India and the Sahelian zone for a time. North Europe, Russia, the Steppes, north China, all should receive more moisture and enjoy improved crops. The Caspian, fed by rivers from the north, should rise.

Drought should plague the U.S. western Great Plains and Canada (says Abbot), and if the traditional pattern holds, we'll have a depression. The world will heave a sigh of relief that the times of strange weather are coming to an end and we're getting back to normal. Average temperatures will still be low, but the patterns will seem familiar.

Alas, as the climatic respite ends, the patterns will shift again and the illusion will be shattered.

It appears that the 20° to 40° latitude band will be favored by the Great Trend. What's in those bands, north and south? These are the areas of the world which the grammar-school geography books refer to as the "temperate" zones.

A great part of the United States and Mexico are in the favored band. Spain, Italy, Greece, Turkey, Morocco, Algeria, Libya, Egypt, Israel, Lebanon, Syria, Jordan, Saudi Arabia, Iraq, Iran, Afghanistan, Pakistan, northern India, Burma, China, Korea, and Japan should all receive increased rainfall in the Northern Hemisphere. In the Southern Hemisphere Chile, Argentina, Paraguay, Uruguay, southern Brazil, Angola, Zambia, Southwest Africa, Malawi, Rhodesia, Botswana, South Africa, Lesotho, Mozambique, the Malagasy Republic,

Australia, and the North Island of New Zealand should all get more rain.

More rain is a mixed blessing. There are places that have been much distressed by increased rain in recent times—e.g., Pakistan and Mexico. Crops suffer from extremes of *any* kind. In general, however, the persistent moisture bodes good.

What's left out of this list?

Lots of big, strong, important, populous countries and regions are left out: France, Germany, Russia, Indonesia, Canada, the Philippines . . .

Arnold Toynbee noted on his birthday in 1974 his belief that the "haves," the nations that are able to support themselves in reasonably comfortable style, will be under constant siege from the "have-nots" as far as he can see into the future. That sounds right to us.

The United States will be in a particularly prominent situation, since it is certain that our immensely sophisticated agricultural system will be able to grow food in quantity, even in times when weather is unpredictable and extreme. Notice that when the Mississippi Valley was flooded in 1973 and vast tracts of cotton-farming land were drowned, the farmers were able to switch to soybeans for the shortened growing season. They brought in a bumper crop of food instead of a bumper crop of fiber. It wasn't really what they had in mind, but it's hard to complain about that kind of performance. Crops are not cheap under these circumstances, but compared with the absence of food, high prices seem attractive.

In a time of great climatic variability, flexibility of response will be a critical factor in survival. U.S. agriculture has already demonstrated its flexibility. It is tempting to attribute this capability of change simply to wealth. We are rich, as a nation, and we can afford to do things that are beyond the resources of people in most other countries. Yet our more important advantage may lie in being uncivilized.

In civilized countries, the primary effort of the bureaucracy is to establish and maintain stability. Nothing is permitted to rock the boat. No risk of any sort is acceptable. Everything is controlled by a complex system that virtually forbids innovation. In such a situation, a belief develops that government ritual almost has magic power to drive away trouble. As long as the rituals are carried out properly, one may expect stability and security.

Inevitably, one trades personal freedom for this illusion of security. Tastes vary. It is obvious that personal freedom is not uniformly valued around the world. We have no editorial complaint to make about this; to each his own.

Still, there is a practical problem associated with belief in government ritual. It is easy to let one's attention wander from the objective of getting necessary work done to the objective of seeing that nothing improper is allowed to happen. A sudden shift from the planting of cotton to the planting of soybeans is terribly upsetting to people whose rituals are all laid out for cotton. By the time they have adjusted to the idea of soybeans, checked the rituals, and given permission to the soybean fanciers, it may be too late to grow the beans. In the United States we retain barely enough disrespect for government ritual to let us survive.

India, to pick an example at random, seems to be more concerned with doing things properly than with growing and distributing food. Americans have been distressed to learn that there is virtually nothing we can do to help India, short of sending food, lots of food, *all* our food. Not just once, but forever. We have not been able to export to India either our technology or our approach to getting a job done. We have tried to send both of these things, but we have failed. Our technology is unacceptable, because it is rather complex and costly. Our approach is unacceptable because it is not Indian. We have been sternly and properly rebuked. We have been told not to meddle in India's affairs, just to send food, more food. Keep sending food.

Americans have been offended by rejection of our assistance when it takes the form of anything but material goods or money. (Especially when we are being cussed out as crude materialists at the same time.) It is a surprise that our self-righteous efforts to get others to develop their own resources with our techniques have been opposed and we have been scolded. Yet it serves us right. If you take a look at the Peace Corps from the viewpoint of many other countries, you can see why that program and others like it have been regarded by some as the most aggressive activity in which the United States has ever engaged.

Consider: We send our smartest, hardest working, and most competent people off to distant lands to live modestly and to create new and useful things out of whatever they find when they get there. Their message is that people are not at the mercy of tradition, not slaves to authority, not dependent on their leaders for important things. Initiative and self-reliance are forcefully demonstrated. While this produces schoolhouses and sanitation systems and power plants and all sorts of things people think they want, it also destroys the local traditions and power structure, scaring people. People hate change, hate new ideas, dread risks. We are still just able to tolerate such things in the United States.

We have felt bad that we cannot persuade some others to help themselves. We felt bad when we ran out of food surplus. We are going to feel worse. We do not have the resources—the land, the supplies, the energy—to feed the world as it stands. Some *will* be left out.

Our tradition has been to help those who are down and out, even when they hate us. Our sympathy for the underdog has always transcended our other feelings and our political leanings. With most of the world down and out, we will face a long series of difficult choices.

Where shall we ship the food we can grow at considerable expense in treasure and labor?

Shall we send it to India and Africa? They need it.

Shall we send it to China and Russia? They need it.

What about England, France, our traditional allies and friends? Those countries are not yet in crisis, but if the trends continue as they have begun, all of the north European countries will be desperate. Cannibalism occurred widely in England during the last century in which tree rings were in their present configuration.

Quite apart from what *should* be done, what *will* be done? We must not confuse morality with reality, lest we enter the field of wishful thinking and abandon the effort to make useful predictions.

Will Americans ration their food to a survival level so that the excess may be shipped elsewhere? Will China? Will Italy? Will Spain?

Will we give food away so that it can equitably be divided among the needy or will we charge for the treasure and labor that we expend to grow that food? Will we charge a profit for the risks we take so we may cover the losses on our failures?

Already we hear some hollering about the nastiness of the "food for crude" attitude which some people have detected in our policies. The opinion has been expressed that Americans would be wicked to ask for petroleum in payment for food. Our possession of food is regarded as pure luck, a gratuitous circumstance that is manifestly unfair to everyone else, unlike the possession of petroleum resources, which had to be earned.

Will we accept that viewpoint and invite the U.N., for example, to take over allocation of food within and outside of the United States? We could certainly get a lot of assistance that way from governments that believe that the United States is wealthy by sheer good luck.

Will we use our capabilities as a weapon to subjugate large parts of the world?

Will we simply close our doors to the world in isolation?

What about Peace on Earth? What are the chances for brotherly love and mutual respect when the mass movements of peoples increase? Will respect for others prevent people from taking drastic action so they don't have to watch their children die? Leave the United States out of the discussion for the moment.

Peru has already increased its territorial boundaries two hundred miles out to sea to prevent other countries from fishing in the waters upon which it depends for livelihood. Many bitter confrontations have occurred over this uncommonly great claim of territorial waters. The Peruvians are especially distressed that the anchovy fisheries have failed because of "overfishing." As a practical matter, it makes no difference what the cause of the failure is—Peru must do *something*, if only to make people realize that the problem is serious. Nobody can tolerate passivity in a crisis.

Iceland, similarly, has extended its territorial claims to discourage other nations from fishing in waters that do not produce enough fish for all. The issue here is food, but the arguments deal chiefly with politics and justice.

The ecologists are going to have an endless series of complaints and will be able to point convincingly to any number of sharp changes in the performance of wildlife. Man-made pollution will take the rap for everything. Serious conservationists, meanwhile, have found it very difficult to determine what is "natural" and what is man-made. While it's true that many measurable changes have occurred in the last few decades, we have not known what to measure them *against*. It's like trying to get accurate measurements of tides in the sea without knowing how much the land is rising and falling at the same time.

Thus far, popular ecological efforts have been limited to loud discussion of individual phenomena and to simple, desirable specific tasks like cleaning up rivers and collecting trash from roadsides.

Ecologists also lean to passing laws against change, risk, new human activity. By and large, when you press an ecologist to tell you what he wants to conserve—ice ages? dinosaurs?—it turns out that he would like to freeze things as they were when he himself was about seven years old. Apart from the general attitude that nothing people do is good, the ecologists are hard pressed to come up with a coherent description of what is happening in the world. Certain anomalies have been baffling.

To use an example close to home again, the elk herds of New Mexico have been increasing unaccountably at a time when the human population of the state is also increasing along with all of the unhappy side effects of "civilization." It is not obvious that "natural" changes in temperature and precipitation have a smaller influence on wildlife than man has.

Again, the United States is in an unprecedented situation. At the very height of our influence in world affairs, we have slowed our population expansion and are approaching a stable population. Additional decrease will give us a *shrinking* population—at the same time that our food-growing capacity is increasing.

The cry of the last years has been that the world is suffering from a population explosion. It seems likely to us that a population *implosion* will mark the end of this century. Famine will account for an appreciable percentage of the world population. Warfare and disease will account for more. Deliberate efforts at birth control may even become effective worldwide in recognition of the difficulties of feeding large numbers of people.

It is probable that anybody who has food will be invited to share it. The invitations may become forceful. It is unlikely that anyone with food will willingly give it up if it means his own starvation. It is unlikely that hungry people will refrain from seizing what they need.

With a shrinking population, the United States will be looked upon as a great place to visit, or better, to move to, or to unite with. Several times in the past other sovereignties have proposed union with the United States.

In the last couple of hundred years initiatives have developed on both sides of the U.S.-Canadian border directed to union of the countries. It is largely a fluke that the union has not taken place during one of the many opportunities. We don't recommend such action (Canada has internal separatist movements already) but one can imagine that a dry, frozen Canada might take the next initiative. Ten years? Fifteen? Considering that one and a half million Mongols conquered three-quarters of a billion people in the last round, might not the union of the United States with Canada be accomplished by Canadian conquest of the United States?

India might wish to share what China grows.

Persia was a great power in times gone by. Might Iran become the breadbasket of Asia, the envy of its neighbors?

Might England, France, and Germany wish to share in Italy's bounty?

Might the Turkish heirs of the Hittites contend again with Egypt?

And might not the barbarians on the borders seize the civilized world again?

Why not?

We may be reading all the signs wrong. The whole structure of related events which we see here may be but an illusion. Even if it is real, perhaps those who disapprove can pass a law against it and, by declaration, make the world a uniformly attractive place for human beings.

And even if the specter cannot be driven off by ritual, we may be cheered by the knowledge that we are here because our ancestors dealt successfully with worse.

Human beings are tough, crafty, and humorous. All human beings are better than whatever is in second place.

It is just possible that pursuit of these ideas can help to save human lives, dignity, and humor.

PUBLISHER'S NOTE

Climate and the Affairs of Men has examined in great detail the mechanisms of changing climate and its historical effects on mankind. But this work is by no means complete. Since the original writing of this book, the scientific community has advanced considerably in its understanding of climatology, and there is much to be brought up to date. Winkless and Browning have been steadily working with new data and believe they have made even more startling discoveries than have been revealed here. Harper's Magazine Press will, therefore, publish the sequel, *Weather, Weapons and Wisdom,* in the spring of 1976.

Bibliography

Abbot, C. G., and Fowle, F. E. 1913. Volcanoes and Climate. Smithsonian Miscellaneous Collections. 60, no. 29.

Aharoni, Yohanan, and Avi-Yonah, Michael. 1968. *The Macmillan Bible Atlas*. New York: Macmillan.

Anderson, R. Y.; Dean, W. E., Jr.; Kirkland, D. W.; and Snider, H. I. 1972. Permian Castile Varved Evaporite Sequence, West Texas and New Mexico. *Geol. Soc. of America Bull.* 83: 59–86.

Aston, W. G., trans. 1971. *Nihongi: Chronicles of Japan from the Earliest Times to A. D. 697*. Rutland, Vt.: Chas. E. Tuttle & Co.

Atlantis and the Searchers. *Newsweek,* July 31, 1967, p. 52.

Atwater, M. A. 1970. Planetary Albedo Changes Due to Aerosols. *Science* 170.

Bakker, E. M. van Zinderen. 1962. A Late-Glacial and Post-Glacial Climatic Correlation Between East Africa and Europe. *Nature* 194.

Baldwin, Ralph B. 1949. *The Face of the Moon*. Chicago: University of Chicago Press.

Bhadawi, M. M., and Ludlow, N. G. T. 1961. Precipitation of Sub-Micron Dust in Still Air by Cloud-Size Water Droplets. *Nature* 190.

Bishai, Wilson B. 1968. *Islamic History of the Middle East*. Boston: Allyn & Bacon, Inc.

Bone, Robert G. 1964. *Ancient History*. Totowa, N. J.: Littlefield, Adams & Co.

Bray, J. R. 1966. Atmospheric Carbon-14 Content During the Past Three Millennia in Relation to Temperature and Solar Activity. *Nature* 209 (5028): 1065–67.

———. 1972. Cyclic Temperature Oscillations from 0–20, 300 Years B.P. *Nature* 237.

Broecker, W. S. 1966. Absolute Dating and the Astronomical Theory of Climatic Change. *Science* 151.

Brooks, C. E. P. 1970. *Climate Through the Ages* (rev. 2d ed.). New York: Dover Publications, Inc.

Brundage, Burr Cartwright. 1972. *A Rain of Darts*. Austin: University of Texas Press.

Bryson, Reid A. 1974. A Perspective on Climatic Change. *Science* 184 (4138).

Cailleux, André. 1968. *Anatomy of the Earth*. New York: McGraw-Hill.

Chao, E. C. T. 1967. Shock Effects in Certain Rock-Forming Minerals. *Science* 156.

Charlson, R. J.; Harrison, H.; and Witt, G. 1972. Aerosol Concentrations: Effect on Planetary Temperatures. *Science* 175.

Cherenkov, V. 1960. A "Saturnian Ring" Around the Earth: Inventor and Efficiency Report. *U.S.S.R.*, no. 2.

Cronin John F. 1971. Recent Volcanism and the Stratosphere. *Science* 172: 847–49.

Dalrymple, G. Brendt; Silver, Eli A.; and Jackson, Everett D. Origin of the Hawaiian Islands. *American Scientist* 61: 294–308.

Dansgaard, W., and Tauber, H. 1969. Glacier Oxygen-18 Content and the Pleistocene Ocean Temperatures. *Science* 166.

Dansgaard et al, K. K. Turekian, ed. 1971. Climatic Record Revealed by the Camp Century Ice Core. *Late Cenozoic Glacial Ages*. New Haven: Yale University Press.

Davidson, Basil. 1964. *The African Past*. An Atlantic Monthly Press Book. Boston: Little, Brown.

———. 1966. *A History of West Africa*. New York: Doubleday Anchor Books.

Davis, Darrell H. 1944. *The Earth and Man—A Human Geography*. New York: Macmillan.

Deetz, James. 1967. *Invitation to Archaeology*. New York: Natural History Press.

Defant, Albert. 1958. *Ebb and Flow: The Tides of Earth, Air, and Water*. Ann Arbor: University of Michigan Press.

Donley, D. L. 1971. Analysis of the Winter Climatic Pattern at the Time of the Mycenaean Decline. Ph.D. dissertation. Madison: University of Wisconsin.

Douglass, A. E. 1919 (vol. 1); 1936 (vol. 3). *Climatic Cycles and Tree Growth*. Carnegie Institution of Washington.

East, W. Gordon. 1967. *The Geography Behind History*. New York: Norton.

Eggenberger, David. 1967. *A Dictionary of Battles from 1479 to the Present*. New York: Thomas Y. Crowell.

Ellis, H. T.; and Pueschel, R. F. 1971. Solar Radiation: Absence of Air Pollution Trends at Mauna Loa. *Science* 172: 845–86.

El Palacio, *Quarterly Journal of the Museum of New Mexico* 78, no. 4.

Elsaesser, Hugh W. 1974. Has Man, Through Increasing Emissions of Particulates, Changed the Climate? Proceedings of Symposium on Atmospheric-Surface Exchange of Particulates and Gaseous Pollutants. Sept. 1974, Richland, Washington.

Emiliani, C. 1972. Quaternary Paleotemperatures and the Duration of the High-Temperature Intervals. *Science* 178.

————. 1972. Interglacial High Sea Levels and the Control of Greenland Ice by the Precession of the Equinoxes. *Science* 178.

Enever, J. E. Mar. 1966. Giant Meteor Impact. *Analog Science Fiction/ Science Fact.*

The Eruption of Krakatoa and Subsequent Phenomena. Report of the Krakatoa Committee of the Royal Society, London, 1888.

Eruptions and Climate: Proof from an Icy Pudding. *New Scientist,* Feb. 17, 1972.

Fagan, Brian M. 1965. *Southern Africa During the Iron Age.* New York: Praeger.

Fessler, Loren. 1963. *China.* Life World Library. New York: Time, Inc.

Flint, Richard Foster. 1971. *Glacial and Quaternary Geology.* New York: Wiley.

Gallant, R. 1963. Changes in the Earth's Axis Due to Large Meteorite Collisions. *Nature* 200.

Gamow, George. 1962. *Gravity.* New York: Doubleday.

Gentner, W.; Glass, B. P.; and Wagner, G. A. 1970. Fission Track Ages and Ages of Deposition of Deep-Sea Microtektites. *Science* 168.

Golovanov, Y. 1960. Can Climate Be Changed? *Priroda* no. 2: 124–28.

Harbottie, Thomas. 1971. *Dictionary of Battles.* New York: Stein & Day.

Harris, A. H.; Schoenwetter, James; and Warren, A. H. 1967. *An Archaeological Survey of the Chuska Valley and the Chaco Plateau, New Mexico,* part 1. Santa Fe: Museum of New Mexico Press.

Hawkes, Jacquetta. 1965. *History of Mankind—Cultural and Scientific Development:* Prehistory, vol. 1, part 1. New York: Mentor/New American Library.

Hawkins, G. S., and White, J. B. 1965. *Stonehenge Decoded.* New York: Dell.

Haynes, C. Vance, Jr. The Earliest Americans. *Science* 166: 709–15.

Heidel, Karen. 1972. Turbidity Trends at Tucson, Arizona. *Science* 177.

Herodotus. 1964. *History of the Greek and Persian War.* New York: Washington Square Press.

Hodge, P. W.; Laulalnen, N.; and Charlson, R. J. 1972. Astronomy and Air Pollution. *Science* 178.

Huntington, Ellsworth. 1924. *Civilization and Climate*. New Haven: Yale University Press.

Huntington, Ellsworth, and Visner, S. S. 1972. *Climatic Changes*. New Haven: Yale University Press.

Jacobs, Melville, and Stern, Bernard J. 1947. *General Anthropology*, 2d ed. New York: Barnes & Noble.

Johnston, M. J., and Mauk, F. J. 1972. Earth Tides and the Triggering of Eruptions from Mt. Stromboli, Italy. *Nature* 239: 266.

Kidder, Alfred V. 1924. *An Introduction to the Study of Southwestern Archaeology with a Preliminary Account of the Excavations at Pecos*. New Haven: Yale University Press.

Knopoff, L. 1964. Earth Tides as Triggering Mechanism for Earthquakes. *Bulletin Seismol. Soc. Amer.* 54.

Ladurie, Emmanuel LeRoy. 1971. *Times of Feast, Times of Famine: A History of Climate Since the Year 1000*. New York. Doubleday.

LaMarche, V. C., Jr. Haleioclimatic Inferences from Long Tree-ring Records. *Science* 183: 1043–48.

Lamb, Harold. 1962. *The Crusades*. New York: Bantam.

———. 1960. *Cyrus the Great*. New York: Doubleday.

Lamb, M. H. The Cooling Effects of Volcanic Outbursts. *Reference Monitor* 52.

Lanning, Edward P. 1967. *Peru Before the Incas*. Englewood Cliffs, N.J.: Prentice-Hall.

Lassetter, R., Jr., and Curtis, D. D. 1936. *Report on Dendrochronological-Hydrological Investigation, Lower Part of the Norris Basin*. Tennessee Valley Authority, Engineering Data Division.

Latham, G., et al. 1971. Moonquakes. *Science* 174.

Leckie, Robert. 1969. *The Wars of America: Quebec to Appomattox*, vol. 1. New York: Bantam.

Lissner, Ivar. 1957. *The Living Past*. New York: Capricorn Books.

Longacre, William A., ed. 1970. *Reconstructing Prehistoric Pueblo Societies*. Albuquerque: University of New Mexico.

Longman, I.˙M. 1959. Formulas for Computing the Tidal Accelerations Due to the Moon and the Sun. *Journal of Geophysical Research* 64, no. 12.

McClelland D.C. 1961. *The Achieving Society*. New York: Van Nostrand.

McEvedy, Colin. 1961. *The Penquin Atlas of Medieval History*. New York: Penguin Books.

McWhirter, N., and McWhirter, R. 1964. *Dunlop Illustrated Encyclopedia of Facts*. New York: Sterling.

Mauk F. J., and Kienle, J. 1973. Microearthquakes at St. Augustine Volcano, Alaska, Triggered by Earth Tides. *Science* 182.

Mauk, F. J., and Johnston, M. J. 1973. *Journal of Geophysical Research* 78.

Meinel, A. B., and Meinel, M. P. 1967. Volcanic Sunset—Glow Stratum: Origin. *Science* 156.

———. 1964. Height of the Glow Stratum from the Eruption of Agung on Bali. *Nature,* Feb. 15.

Milliman, John D., and Emery, K. O. Sea Levels During the Past 35,000 Years. *Science* 164: 177–79.

Mullan, D. J. 1973. *Earthquake Waves and the Geomagnetic Dynamo. Science* 181.

O'Keefe, John A., ed. 1963. *Tektites.* Chicago: University of Chicago Press.

Peirce, Jodie. 1974. Cultural Sensitivity to Environmental Change: 1816, the Year Without a Summer. Paper for N.S.F. University of Wisconsin.

Phillips, E. D. 1965. *The Royal Hordes.* New York: McGraw-Hill.

Ramp Mining for Nickel. 1968. *Nickel Topics* 21, no. 6.

Rasool, S. J., and Schneider, S. H. 1971. Atmospheric Carbon Dioxide and Aerosols: Effects of Large Increases on Global Climate. *Science* 173.

Renfrew. Colin. 1971. Carbon 14 and the Prehistory of Europe. *Scientific American* 225, no. 4.

Richardson, L. F. 1960. *The Statistics of Deadly Quarrels.* New York: Quadrangle.

Rinehart, John S. 1972. 18.6-Year Earth Tide Regulates Geyser Activity. *Science* 177.

Roosen, R. G.; Angione, R. J.; and Klemcke, C. H. Worldwide Variations in Atmospheric Transmission, I. Baseline Results from Smithsonian Observatory. Unpublished paper.

Rowlett, Ralph M. The Iron Age North of the Alps. *Science* 161: 123–34.

Schulman, Edmund. 1956. *Dendroclimatic Changes in Semiarid America.* Tucson: University of Arizona Press.

Schurz, William L. 1964. *This New World: The Civilization of Latin America.* New York: Dutton.

Shaw, H. R. 1970. Earth Tides, Global Heat Flow, and Tectonics. *Science* 168.

Silverberg, Robert. 1966. *Empires in the Dust.* New York: Bantam.

Simpson, L. B. 1971. *Many Mexicos.* Berkeley: University of California Press.

Smith, Huston. 1958. *The Religions of Man.* New York: Harper & Row.

Stipp, J. L.; Hollister, C. W.; and Dirrim, A. W. 1967. *The Rise and Development of Western Civilization to 1660,* vol. 1. New York: Wiley.

Stoiber, R. B., and Jepsen, A. 1973. Sulfur Dioxide Contributions to the Atmosphere by Volcanoes. *Science* 189: 577–78.

Suess, Hans E. 1965. Secular Variations of the Cosmic-Ray-Produced

Carbon 14 in the Atmosphere and Their Interpretations. *Journal of Geophysical Research* 70.

Tannehill, I. R. 1947. *Drought: Its Causes and Effects*. Princeton: Princeton University Press.

Taylor, P. S., and Stoiber, R. E. 1973. *Geol. Soc. Amer. Bull.* 84.

Thucydides. 1954. *The Peloponnesian War*. New York: Penguin Books.

Treistman, Judith M. China at 1000 B.C.: A Cultural Mosaic. *Science* 160: 853–56.

———. 1972. *The Prehistory of China: An Archaeological Exploration*. New York: Doubleday.

Van Houten, F. B. 1969. Mollasse Facies: Records of Worldwide Stress. *Science* 166: 1506–08.

Veeh, H. H., and Chappell, J. 1970. Astronomical Theory of Climatic Change: Support from New Guinea. *Science* 167.

Von Hagen, Victor W. 1958. *The Aztec: Man and Tribe*. New York: Mentor/New American Library.

———. 1957. *Realm of the Incas*. New York: Mentor/New American Library.

———. 1960. *World of the Maya*. New York: Mentor/New American Library.

Vonnegut, B.; McConnell, R. K., Jr.; and Allen, R. V. 1966. Evaporation of Lava and Its Condensation from the Vapour Phase in Terrestrial and Lunar Volcanism. *Nature*, Jan. 29.

Watson, F. C. 1962. *Between the Planets*. New York: Doubleday.

Whitcomb, J. H.; Garmany, J. L.; and Anderson, D. L. 1973. Earthquake Prediction: Variation of Seismic Velocities before the San Fernando Earthquake. *Science* 180: 632–35.

Wilson, J. Tuzo. 1972. Mao's Almanac. 3000 Years of Killer Earthquakes. *Saturday Review*, Feb. 19, 1972.

Woolley, Sir Leonard. 1963. *History of Mankind: Cultural and Scientific Development: The Beginnings of Civilization*, vol. 1, part 2. New York: Mentor/New American Library.

Wormington, H. M. 1947. *Prehistoric Indians of the Southwest*. Denver Museum of Natural History.

Wren, M. C., ed. 1968. *The Course of Russian History*, 3d. ed. New York: Macmillan.

Wright, Quincy. 1942. *A Study of War*. Chicago: University of Chicago Press.

Zotkin, I. T. The Tunguska Meteorite. Soviet Academy of Sciences. *Meteoritika* 20: 40–53.